Contents

How to use this guide

This guide is divided into two parts:

- **Part 1** Explains what fire risk assessment is and how you might go about it. Fire risk assessment should be the foundation for all the fire precautions in your premises.
- **Part 2** Provides further guidance on fire precautions. The information is provided for you and others to dip into during your fire risk assessment or when you are reviewing your precautions.

The appendices provide example checklists, some detailed technical information on fire-resisting elements and advice on historic buildings.

This guide is one of a series of guides listed on the back cover.

The rest of this introduction explains how the law applies.

Technical terms are explained in the glossary and references to other publications listed at the end of the publication are identified by a superscript number in the text.

In this guide reference is made to British Standards and standards provided by other bodies. The standards referred to are intended for guidance only and other standards could be used. Reference to any particular standard is not intended to confer a presumption of conformity with the requirements of the Regulatory Reform (Fire Safety) Order 2005 (the Order)[1].

The level of necessary safety (or service) must be dictated by the findings of your risk assessment so you may need to do more or less than that specified in any particular standard referred to. You must be prepared to show that what you have done complies with any requirements or prohibitions of the Order[1] irrespective of whether you have relied on a particular standard.

A full list of references, e.g.[1], can be found at the back of this book.

Preface

This guidance gives advice on how to avoid fires and how to ensure people's safety if a fire does start. Why should you read it? Because:

- Fire kills. In 2004 (England and Wales) fire and rescue services attended over 33,400 fires in non-domestic buildings. These fires killed 38 people and injured over 1,300.
- Fire costs money. The costs of a serious fire can be high and afterwards many businesses do not reopen. In 2004, the costs as a consequence of fire, including property damage, human casualties and lost business, were estimated at £2.5 billion.

This guide applies to England and Wales only. It does not set prescriptive standards, but provides recommendations and guidance for use when assessing the adequacy of fire precautions in offices and shops. Other fire risk assessment methods may be equally valid to comply with fire safety law. The guide also provides recommendations for the fire safety management of the premises.

Your existing fire safety arrangements may not be the same as the recommendations used in this guide but, as long as you can demonstrate that they meet an equivalent standard of fire safety, they are likely to be acceptable. If you decide that your existing arrangements are not satisfactory there may be other ways to comply with fire safety law. This means there is no obligation to adopt any particular solution in this guide if you prefer to meet the relevant requirement in some other way.

Where the building has been recently constructed or significantly altered, the fire detection and warning arrangements, escape routes and facilities for the fire and rescue service should have been designed, constructed and installed in accordance with current building regulations. In such cases, it is likely that these measures will be satisfactory as long as they are being properly maintained and no significant increase in risk has been introduced.

In addition, earlier legislation, including local authorities acts may have imposed additional requirements for fire precautions in large offices and shops.

This guide should not be used to design fire safety in new buildings. Where alterations are proposed to existing offices and shops, they may be subject to building regulations. However, it can be used to develop the fire safety strategy for the building.

Introduction

WHO SHOULD USE THIS GUIDE?

This guide is for all employers, managers, occupiers and owners of offices and shops. Details of other guides in the series are listed on the back cover. It tells you about how you might comply with fire safety law, helps you to carry out a fire risk assessment and identify the general fire precautions you need to have in place.

This guide is intended for premises where the main use of the building or part of the building is an office or shop. It may also be suitable for the individual units within other complexes, although consultation with the other managers will be necessary as part of an integrated risk assessment for the complex.

Also, where you handle and store flammable materials and substances, it will help you take account of these in your risk assessment and help you determine the necessary precautions to take to minimise the likelihood of them being involved in a fire.

It has been written to provide guidance for a responsible person, to help them to carry out a fire risk assessment in less complex offices and shops. If you read the guide and decide that you are unable to apply the guidance, then you should seek expert advice from a competent person. More complex premises will probably need to be assessed by a person who has comprehensive training or experience in fire risk assessment. However this guide can be used for multi-occupied buildings to address fire safety issues within their individual occupancies.

It may also be useful for:

- employees;
- employee-elected representatives;
- trade union-appointed health and safety representatives;
- enforcing authorities; and
- all other people who have a role in ensuring fire safety in offices and shops.

If your premises are listed as of historic interest, also see Appendix C.

Fire safety is only one of many safety issues with which management must concern themselves to minimise the risk of injury or death to staff or the public. Unlike most of the other safety concerns, fire has the potential to injure or kill large numbers of people very quickly. This guidance is concerned only with fire safety but many of the measures discussed here will impact upon other safety issues, and vice versa. It is recognised that these various differing safety demands can sometimes affect one another and managerment should consult other interested agencies (e.g. Health & Safety Executive (HSE)) where necessary to confirm that they are not contravening other legislation/guidance.

You can get advice about minimising fire losses from your insurer.

THE FIRE SAFETY ORDER

Previous general fire safety legislation

The Order[1] replaces previous fire safety legislation. Any fire certificate issued under the Fire Precautions Act 1971[2] will cease to have any effect. If a fire certificate has been issued in respect of your premises or the premises were built to recent building regulations, as long as you have made no material alterations and all the physical fire precautions have been properly maintained, then it is unlikely you will need to make any significant improvements to your existing physical fire protection arrangements to comply with the Order.[1] However, you must still carry out a fire risk assessment and keep it up to date to ensure that all the fire precautions in your premises remain current and adequate.

If you have previously carried out a fire risk assessment under the Fire Precautions (Workplace) Regulations 1997,[3] as amended 1999,[4] and this assessment has been regularly reviewed then all you will need to do now is revise that assessment taking account of the wider scope of the Order[1] as described in this guide.

Your premises may also be subject to the provisions of a licence or registration (e.g. under the Licensing Act 2003[78]) of the case, and the fire authority may wish to review your risk assessment as part of the licensing approval process. Fire safety conditions within your licence should not be set by a licensing authority where the Order[1] applies.

Background

The Order[1] applies in England and Wales. It covers general fire precautions and other fire safety duties which are needed to protect 'relevant persons' in case of fire in and around most 'premises'. The Order requires fire precautions to be put in place 'where necessary' and to the extent that it is reasonable and practicable in the circumstances of the case.

Responsibility for complying with the Order[1] rests with the 'responsible person'. In a workplace, this is the employer and any other person who may have control of any part of the premises, e.g. the occupier or owner. In all other premises the person or people in control of the premises will be responsible. If there is more than one responsible person in any type of premises (e.g. a multi-occupied complex), all must take all reasonable steps to co-operate and co-ordinate with each other.

If you are the responsible person you must carry out a fire risk assessment which must focus on the safety in case of fire of all 'relevant persons'. It should pay particular attention to those at special risk, such as disabled people, those who you know have special needs and young persons and must include consideration of any dangerous substance liable to be on the premises. Your fire risk assessment will help you identify risks that can be removed or reduced and to decide the nature and extent of the general fire precautions you need to take.

If your organisation employs five or more people, your premises are licensed or an alterations notice is in force, you must record the significant findings of the assessment. It is good practice to record your significant findings in any case.

There are some other fire safety duties you need to comply with:

- **You must** appoint one or more competent persons, depending on the size and use of your premises, to carry out any of the preventive and protective measures required by the Order[1] (you can nominate yourself for this purpose). A competent person is someone with enough training and experience or knowledge and other qualities to be able to implement these measures properly.
- **You must** provide your employees with clear and relevant information on the risks to them identified by the fire risk assessment, about the measures you have taken to prevent fires, and how these measures will protect them if a fire breaks out.
- **You must** consult your employees (or their elected representatives) about nominating people to carry out particular roles in connection with fire safety and about proposals for improving the fire precautions.
- **You must**, before you employ a child, provide a parent with clear and relevant information on the risks to that child identified by the risk assessment, the measures you have put in place to prevent/protect them from fire and inform any other responsible person of any risks to that child arising from their undertaking.
- **You must** inform non-employees, such as temporary or contract workers, of the relevant risks to them, and provide them with information about who are the nominated competent persons, and about the fire safety procedures for the premises.
- **You must** co-operate and co-ordinate with other responsible persons who also have premises in the building, inform them of any significant risks you find and how you will seek to reduce/control those risks which might affect the safety of their employees.
- **You must** provide the employer of any person from an outside organisation who is working in your premises (e.g. an agency providing temporary staff) with clear and relevant information on the risks to those employees and the preventive and protective measures taken. You must also provide those employees with appropriate instructions and relevant information about the risks to them.
- If you are not the employer but have any control of premises which contain more than one workplace, **you are also responsible** for ensuring that the requirements of the Order[1] are complied with in those parts over which you have control.
- **You must** consider the presence of any dangerous substances and the risk this presents to relevant persons from fire.
- **You must** establish a suitable means of contacting the emergency services and provide them with any relevant information about dangerous substances.

- **You must** provide appropriate information, instruction and training to your employees, during their normal working hours, about the fire precautions in your workplace, when they start working for you, and from time to time throughout the period they work for you.
- **You must** ensure that the premises and any equipment provided in connection with firefighting, fire detection and warning, or emergency routes and exits are covered by a suitable system of maintenance and are maintained by a competent person in an efficient state, in efficient working order and in good repair.
- **Your employees must** co-operate with you to ensure the workplace is safe from fire and its effects, and must not do anything that will place themselves or other people at risk.

The above examples outline some of the main requirements of the Order.[1] The rest of this guide will explain how you might meet these requirements.

Who enforces the Fire Safety Order?

The local fire and rescue authority (the fire and rescue service) will enforce the Order[1] in most premises. The exceptions are:

- Crown-occupied/owned premises where Crown fire inspectors will enforce;
- premises within armed forces establishments where the defence fire and rescue service will enforce;
- certain specialist premises including construction sites, ships (under repair or construction) and nuclear installations, where the HSE will enforce; and
- sports grounds and stands designated as needing a safety certificate by the local authority, where the local authority will enforce.

The enforcing authority will have the power to inspect your premises to check that you are complying with your duties under the Order.[1] They will look for evidence that you have carried out a suitable fire risk assessment and acted upon the significant findings of that assessment. If you are required to record the outcome of the assessment they will expect to see a copy.

If the enforcing authority is dissatisfied with the outcome of your fire risk assessment or the action you have taken, they may issue an enforcement notice that requires you to make certain improvements or, in extreme cases, a prohibition notice that restricts the use of all or part of your premises until improvements are made.

If your premises are considered by the enforcing authority to be or have potential to be high risk, they may issue an alterations notice that requires you to inform them before you make any changes to your premises or the way they are used.

Failure to comply with any duty imposed by the Order[1] or any notice issued by the enforcing authority is an offence. You have a right of appeal to a magistrates court against any notice issued. Where you agree that there is a need for improvements to your fire precautions but disagree with the enforcing authority on the technical solution to be used (e.g. what type of fire alarm system is needed) you may agree to refer this for independent determination.

If having read this guide you are in any doubt about how fire safety law applies to you, contact the fire safety office at your local fire and rescue service.

If your premises were in use before 2006, then they may have been subject to the Fire Precautions Act[2] and the Fire Precautions (Workplace) Regulations.[3,4] Where the layout (means of escape) and other fire precautions have been assessed by the fire and rescue service to satisfy the guidance that was then current, then it is likely that your premises already conform to many of the recommendations here, providing you have undertaken a fire risk assessment as required by the Fire Precautions (Workplace) Regulations.[3,4]

New buildings or significant building alterations should be designed to satisfy current building regulations[24] (which address fire precautions).

However, you will still need to carry out a fire risk assessment, or review your existing assessment (and act on your findings), to comply with the Order.[1]

Part 1 Fire risk assessment

MANAGING FIRE SAFETY

Good management of fire safety is essential to ensure that fires are unlikely to occur; that if they do occur they are likely to be controlled or contained quickly, effectively and safely; or that, if a fire does occur and grow, everyone in your premises is able to escape to a place of total safety easily and quickly.

The risk assessment that you must carry out will help you ensure that your fire safety procedures, fire prevention measures, and fire precautions (plans, systems and equipment) are all in place and working properly, and the risk assessment should identify any issues that need attention. Further information on managing fire safety is available in Part 2 on page 39.

WHAT IS A FIRE RISK ASSESSMENT?

A fire risk assessment is an organised and methodical look at your premises, the activities carried on there and the likelihood that a fire could start and cause harm to those in and around the premises.

The aims of the fire risk assessment are:

- To identify the fire hazards.
- To reduce the risk of those hazards causing harm to as low as reasonably practicable.
- To decide what physical fire precautions and management arrangements are necessary to ensure the safety of people in your premises if a fire does start.

The term 'where necessary' (see Glossary) is used in the Order,[1] therefore when deciding what fire precautions and management arrangements are necessary you will need to take account of this definition.

The terms 'hazard' and 'risk' are used throughout this guide and it is important that you have a clear understanding of how these should be used.

- **Hazard:** anything that has the potential to cause harm.
- **Risk:** the chance of that harm occurring.

If your organisation employs five or more people, or your premises are licensed or an alterations notice requiring it is in force, then the significant findings of the fire risk assessment, the actions to be taken as a result of the assessment and details of anyone especially at risk must be recorded. You will probably find it helpful to keep a record of the significant findings of your fire risk assessment even if you are not required to do so.

HOW DO YOU CARRY OUT A FIRE RISK ASSESSMENT?

A fire risk assessment will help you determine the chances of a fire starting and the dangers from fire that your premises present for the people who use them and any person in the immediate vicinity. The assessment method suggested in this guide shares the same approach as that used in general health and safety legislation and can be carried out either as part of a more general risk assessment or as a separate exercise. As you move through the steps there are checklists to help you.

Before you start your fire risk assessment, take time to prepare, and read through the rest of Part 1 of this guide.

Much of the information for your fire risk assessment will come from the knowledge your employees, colleagues and representatives have of the premises, as well as information given to you by people who have responsibility for other parts of the building. A tour of your premises will probably be needed to confirm, amend or add detail to your initial views.

It is important that you carry out your fire risk assessment in a practical and systematic way and that you allocate enough time to do a proper job. It must take the whole of your premises into account, including outdoor locations and any rooms and areas that are rarely used. If your premises are small you may be able to assess them as a whole. In larger premises you may find it helpful to divide them into rooms or a series of assessment areas using natural boundaries, e.g. process areas (such as bakeries and cooking facilities in shops), offices, stores, as well as corridors, stairways and external routes.

If your premises are in a multi-use complex then the information on hazard and risk reduction will still be applicable to you. However, any alterations to the use or structure of your individual unit will need to take account of the overall fire safety arrangements in the building.

Your premises may be simple, with few people present or with a limited degree of business activity, but if it forms part of a building with different occupancies, then the measures provided by other occupiers may have a direct effect on the adequacy of the fire safety measures in your premises.

Under health and safety law (enforced by the HSE or the local authority) you are required to carry out a risk assessment in respect of any work processes in your workplace and to take or observe appropriate special, technical or organisational measures. If your health and safety risk assessment identifies that these processes are likely to involve the risk of fire or the spread of fire then you will need to take this into account during your fire risk assessment under the Order,[1] and prioritise actions based on the level of risk.

You need to appoint one or more competent persons (this could be you) to carry out any of the preventive and protective measures needed to comply with the Order.[1] This person could be you, or an appropriately trained, employee or, where appropriate, a third party.

Your fire risk assessment should demonstrate that, as far as is reasonable, you have considered the needs of all relevant persons, including disabled people.

Figure 1 shows the five steps you need to take to carry out a fire risk assessment.

FIRE SAFETY RISK ASSESSMENT

1 Identify fire hazards

Identify:

Sources of ignition
Sources of fuel
Sources of oxygen

2 Identify people at risk

Identify:

People in and around the premises
People especially at risk

3 Evaluate, remove, reduce and protect from risk

Evaluate the risk of a fire occurring
Evaluate the risk to people from fire
Remove or reduce fire hazards
Remove or reduce the risks to people

- Detection and warning
- Fire-fighting
- Escape routes
- Lighting
- Signs and notices
- Maintenance

4 Record, plan, inform, instruct and train

Record significant finding and action taken
Prepare an emergency plan
Inform and instruct relevant people; co-operate and co-ordinate with others
Provide training

5 Review

Keep assessment under review
Revise where necessary

Remember to keep to your fire risk assessment under review.

Figure 1: The five steps of a fire risk assessment

STEP 1 IDENTIFYING FIRE HAZARDS

For a fire to start, three things are needed:

- a source of ignition;
- fuel; and
- oxygen.

If any one of these is missing, a fire cannot start. Taking measures to avoid the three coming together will therefore reduce the chances of a fire occurring.

The remainder of this step will advise on how to identify potential ignition sources, the materials that might fuel a fire and the oxygen supplies that will help it burn.

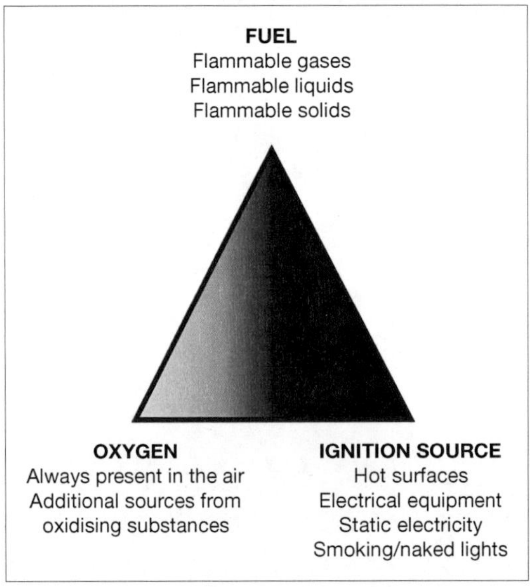

Figure 2: The fire triangle

1.1 Identify sources of ignition

You can identify the potential ignition sources in your premises by looking for possible sources of heat which could get hot enough to ignite material found in your premises. These sources could include:

- smokers' material, e.g. cigarettes, matches and lighters;
- naked flames, e.g. candles or gas or liquid-fuelled open-flame equipment;
- electrical, gas or oil-fired heaters (fixed or portable);
- hot processes, e.g. welding by contractors or shrink wrapping;
- cooking equipment;
- faulty or misused electrical equipment;
- lighting equipment, e.g. halogen lamps or display lighting too close to stored products;
- hot surfaces and obstruction of equipment ventilation, e.g. office equipment; and
- arson.

Indications of 'near-misses', such as scorch marks on furniture or fittings, discoloured or charred electrical plugs and sockets, cigarette burns etc., can help you identify hazards which you may not otherwise notice.

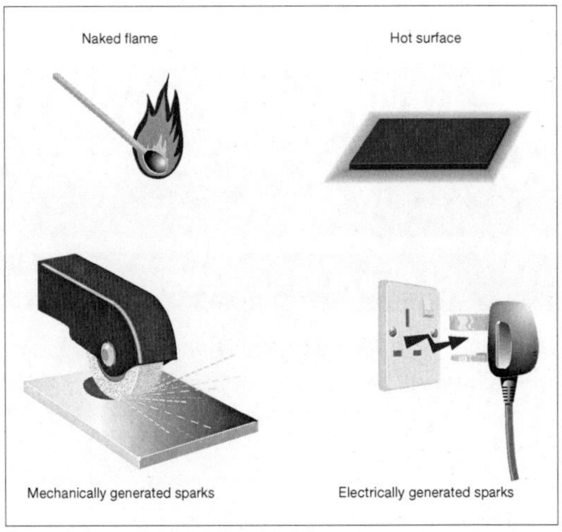

Figure 3: Sources of ignition

1.2 Identify sources of fuel

Anything that burns is fuel for a fire. You need to look for the things that will burn reasonably easily and are in enough quantity to provide fuel for a fire or cause it to spread to another fuel source. Some of the most common 'fuels' found in offices and shops are:

- flammable-liquid-based products, such as paints, varnishes, thinners and adhesives;
- flammable liquids and solvents, such as white spirit, methylated spirit, cooking oils and disposable cigarette lighters;
- flammable chemicals, such as certain cleaning products, photocopier chemicals and dry cleaning that uses hydrocarbon solvents;
- packaging materials, stationery, advertising material and decorations;
- plastics and rubber, such as video tapes, polyurethane foam-filled furniture and polystyrene-based display materials;
- textiles and soft furnishings, such as hanging curtains and clothing displays;
- waste products, particularly finely divided items such as shredded paper and wood shavings, off cuts, and dust; and
- flammable gases such as liquefied petroleum gas (LPG).

You should also consider the materials used to line walls and ceilings, e.g. polystyrene or carpet tiles, the fixtures and fittings, and how they might contribute to the spread of fire. Further information is available in Part 2, Section 1.

1.3 Identify sources of oxygen

The main source of oxygen for a fire is in the air around us. In an enclosed building this is provided by the ventilation system in use. This generally falls into one of two categories: natural airflow through doors, windows and other openings; or mechanical air conditioning systems and air handling systems. In many buildings there will be a combination of systems, which will be capable of introducing/extracting air to and from the building.

Figure 4: Label on oxidising materials

Additional sources of oxygen can sometimes be found in materials used or stored at premises such as:

- some chemicals (oxidising materials), which can provide a fire with additional oxygen and so help it burn. These chemicals should be identified on their container (and Control of Substances Hazardous to Health data sheet, see Figure 4) by the manufacturer or supplier who can advise as to their safe use and storage;

- oxygen supplies from cylinder storage and piped systems, e.g. oxygen used in welding processes; and
- pyrotechnics (fireworks), which contain oxidising materials and need to be treated with great care.

Checklist

- Have you identified all potential ignition sources? ☐
- Have you identified all potential fuel sources? ☐
- Have you identified all potential sources of oxygen? ☐
- Have you made a note of your findings? ☐

STEP 2 IDENTIFYING PEOPLE AT RISK

As part of your fire risk assessment, you need to identify those at risk if there is a fire. To do this you need to identify where you have people working, either at permanent workstations or at occasional locations around the premises, and to consider who else may be at risk, such as customers, visiting contractors etc., and where these people are likely to be found.

You must consider all the people who use the premsies but you should pay particular attention to people who may be especially at risk such as:

- employees who work alone and/or in isolated areas, e.g. cleaners, security staff;
- people who are unfamiliar with the premises, e.g. seasonal workers, contractors, visitors and customers;
- people with disabilities* or those who may have some other reason for not being able to leave the premises quickly, e.g. elderly customers or parents with children;
- other persons in the immediate vicinity of the premises; and
- people with language difficulties.

In evaluating the risk to people with disabilities you may need to discuss their individual needs with them. In larger buildings used extensively by the public you may need to seek professional advice.

Further guidance on people with special needs is given in Part 2, Section 1.

Checklist

- Have you identified who is at risk? ☐
- Have you identified why they are at risk? ☐
- Have you made a note of your findings? ☐

* Visit the Disability Rights Commission website on www.drc-gb.org for more information.

STEP 3 EVALUATE, REMOVE, REDUCE AND PROTECT FROM RISK

The management of the premises and the way people use it will have an effect on your evaluation of risk. Management may be your responsibility alone or there may be others, such as the building owners or managing agents, who also have responsibilities. In multi-occupied buildings all those with some control must co-operate and you need to consider the risk generated by others in the building.

3.1 Evaluate the risk of a fire occuring

The chances of a fire starting will be low if your premises has few ignition sources and combustible materials are kept away from them.

In general, fires start in one of three ways:

- accidentally, such as when smoking materials are not properly extinguished or when lighting displays are knocked over;
- by act or omission, such as when electrical office equipment is not properly maintained, or when waste packaging is allowed to accumulate near to a heat source; and
- deliberately, such as an arson attack involving setting fire to external rubbish bins placed too close to the building.

Look critically at your premises and try to identify any accidents waiting to happen and any acts or omissions which might allow a fire to start. You should also look for any situation that may present an opportunity for an arsonist

Further guidance in Part 2, Section 1 on evaluating the risk of a fire starting.

3.2 Evaluate the risk to people

In Step 2 you identified the people likely to be at risk should a fire start anywhere in the premises and earlier in Step 3 you identified the chances of a fire occurring. It is unlikely that you will have concluded that there is no chance of a fire starting anywhere in your premises so you now need to evaluate the actual risk to those people should a fire start and spread from the various locations that you have identified.

While determining the possible incidents, you should also consider the likelihood of any particular incident; but be aware that some very unlikely incidents can put many people at risk.

To evaluate the risk to people in your premises, you will need to understand the way fire can spread. Fire is spread by three methods:

- convection;
- conduction; and
- radiation.

Convection

Fire spread by convection is the most dangerous and causes the largest number of injuries and deaths. When fires start in enclosed spaces such as buildings, the smoke rising from the fire gets trapped by the ceiling and then spreads in all directions to form an ever-deepening layer over the entire room space. The smoke will pass through any holes or gaps in the walls, ceiling and floor into other parts of the building. The heat from the fire gets trapped in the building and the temperature rises.

Figure 5: Smoke moving through a building

Conduction

Some materials, such as metal shutters and ducting, can absorb heat and transmit it to the next room, where it can set fire to combustible items that are in contact with the heated material.

Radiation

Radiation heats the air in the same way as an electric bar heater heats a room. Any material close to a fire will absorb the heat until the item starts to smoulder and then burn.

Smoke produced by a fire also contains toxic gases which are harmful to people. A fire in a building with modern fittings and materials generates smoke that is thick and black, obscures vision, causes great difficulty in breathing and can block the escape routes.

It is essential that the means of escape and other fire precautions are adequate to ensure that everyone can make their escape to a place of total safety before the fire and its effects can trap them in the building.

In evaluating this risk to people you will need to consider situations such as:

- fire starting on a lower floor affecting the only escape route for people on upper floors or the only escape route for people with disabilities;
- fire developing in an unoccupied space that people have to pass by to escape from the building;
- fire or smoke spreading through a building via routes such as vertical shafts, service ducts, ventilation systems, poorly installed, poorly maintained or damaged walls, partitions and ceilings affecting people in remote areas;
- fire starting in a service room and affecting hazardous materials;
- fire spreading rapidly through the building because of combustible structural elements and/or large quantities of combustible goods;
- rapid vertical fire spread in racked displays;

- fire and smoke spreading through a building due to poor installation of fire precautions, e.g. incorrectly installed fire doors (see Appendix B2 for more information on fire doors) or incorrectly installed services penetrating fire walls; and
- fire and smoke spreading through the building due to poorly maintained and damaged fire doors or fire doors being wedged open.

Further guidance on fire risks is given in Part 2, Section 1.

3.3 Remove or reduce the hazards

Having identified the fire hazards in Step 1, you now need to remove those hazards if reasonably practicable to do so. If you cannot remove the hazards, you need to take reasonable steps to reduce them if you can. This is an essential part of fire risk assessment and as a priority this must take place before any other actions.

Ensure that any actions you take to remove or reduce fire hazards or risk are not substituted by other hazards or risks. For example, if you replace a flammable substance with a toxic or corrosive one, you must consider whether this might cause harm to people in other ways.

Remove or reduce sources of ignition

There are various ways that you can reduce the risk caused by potential sources of ignition, for example:

- Wherever possible replace a potential ignition source by a safer alternative.
- Replace naked flame and radiant heaters with fixed convector heaters or a central heating system. Restrict the movement of and guard portable heating appliances.
- Seperate ignition hazards and combustibles e.g. ensure sufficient clear space between lights and combustibles.
- Operate a safe smoking policy in designated smoking areas and prohibit smoking elsewhere.
- Ensure electrical and mechanical and gas equipment is installed, used, maintained and protected in accordance with the manufacturer's instructions.
- Check all areas where hot work (e.g. welding) has been carried out to ensure that no ignition has taken place or any smouldering materials remain that may cause of fire.
- Ensure that no-one carrying out work on gas fittings which involves exposing pipes that contain or have contained flammable gas uses any source of ignition such as blow-lamps or hot-air guns.
- Take precautions to avoid arson.

Remove or reduce sources of fuel

There are various ways that you can reduce the risks caused by materials and substances which burn, for example:

- Reduce stocks of flammable materials, liquids and gases on display in public areas to a minimum. Keep remaining stock in dedicated storerooms or storage areas where the public are not allowed to go, and keep the minimum required for the operation of the business.

- Ensure flammable materials, liquids and gases, are kept to a minimum, and are stored properly with adequate separation distances between them.
- Keep areas containing flammable gasses ventilated.
- Do not keep flammable solids, liquids and gases together.
- Remove, or treat large areas of highly combustible wall and ceiling linings, e.g. polystyrene or carpet tiles, to reduce the rate of flame spread across the surface.
- Develop a formal system for the control of combustible waste by ensuring that waste materials and rubbish are not allowed to build up and are carefully stored until properly disposed of, particularly at the end of the day.
- Take action to avoid storage areas being vulnerable to arson or vandalism.
- Check all areas where hot work (e.g. welding) has been carried out to ensure that no ignition has taken place and no smouldering or hot materials remain that may cause a fire later.

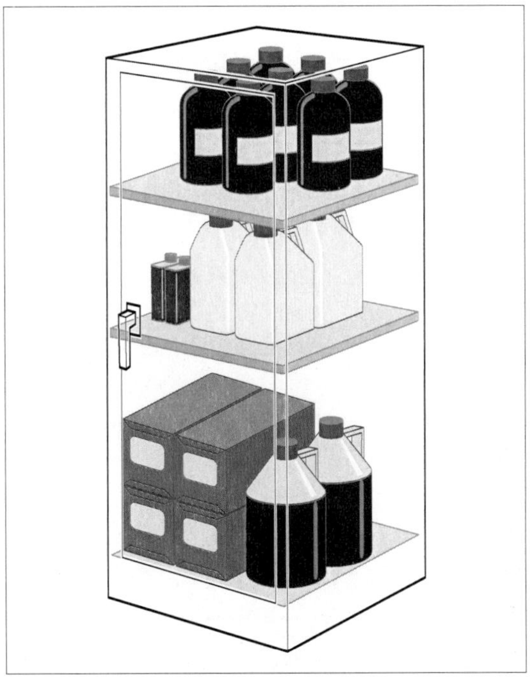

Figure 6: Storage of flammables

Further guidance on removing and reducing hazards is given in Part 2, Section 1.

Remove or reduce sources of oxygen

You can reduce the potential source of oxygen supplied to a fire by:

- closing all doors, windows and other openings not required for ventilation, particularly out of working hours;
- shutting down ventilation systems which are not essential to the function of the premises;
- not storing oxidising materials near or with any heat source or flammable materials; and
- controlling the use and storage of oxygen cylinders, ensuring that they are not leaking, are not used to 'sweeten' the atmosphere, and that where they are located is adequately ventilated.

3.4 Remove or reduce the risks to people

Having evaluated and addressed the risk of fire occuring and the risk to people (preventative measures) it is unlikely that you will be able to conclude that no risk remains of fire starting and presenting a risk to people in your premises.

You now need to reduce any remaining fire risk to people to as low as reasonably practicable, by ensuring that adequate fire precautions are in place to warn people in the event of a fire and allow them to safely escape.

The rest of this step describes the fire protection measures you may wish to adopt to reduce the remaining fire risk to people (see Steps 3.4.1 to 3.4.6).

The level of fire protection you need to provide will depend on the level of risk that remains in the premises after you have removed or reduced the hazards and risks. Part 2, Section 4.1 can help you decide the level of risk that you may still have.

Flexibility of fire protection measures

Flexibility will be required when applying this guidance, the level of fire protection should be proportional to the risk posed to the safety of the people in the premises. Therefore, the objective should be to reduce the remaining risk to a level as low as reasonably practicable. The higher the risk of fire and risk to life, the higher the standards of fire protection will need to be.

Your premises may not exactly fit the solutions suggested in this guide and they may need to be applied in a flexible manner without compromising the safety of the occupants.

For example, if the travel distance is in excess of the norm for the level of risk you have determined, (see Part 2, Table 2 on page 68), it may be necessary to do any one or a combination of the following to compensate:

- Provide earlier warning of fire using automatic fire detection.
- Revise the layout to reduce travel distances.
- Reduce the fire risk by removing or reducing combustible materials and/or ignition sources.
- Control the number of people in the premises.
- Limit the area to trained staff only (no public).
- Increase staff training and awareness.

Note: The above list is not exhaustive and is only used to illustrate some examples of trade-offs to provide safe premises.

If you decide to significantly vary away from the benchmarks in this guidance then you should seek expert advice before doing so.

3.4.1 Fire-detection and warning systems

In some small, open-plan, single-storey offices and shops, a fire may be obvious to everyone as soon as it starts. In these cases, where the number and position of exits and the travel distance to them is adequate, a simple shout of 'fire' or a simple manually operated device, such as a gong or air horn that can be heard by everybody when operated from any single point within the building, may be all that is needed. Where a simple shout or manually operated device is not adequate, it is likely that an electrical fire warning system will be required.

In larger premises, particularly those with more than one floor, where an alarm given from any single point is unlikely to be heard throughout the building an electrical system incorporating sounders and manually operated call points (break-glass boxes) is likely to be required. This type of system is likely to be acceptable where all parts of the building are occupied at the same time and it is unlikely that a fire could start without somebody noticing it quickly.

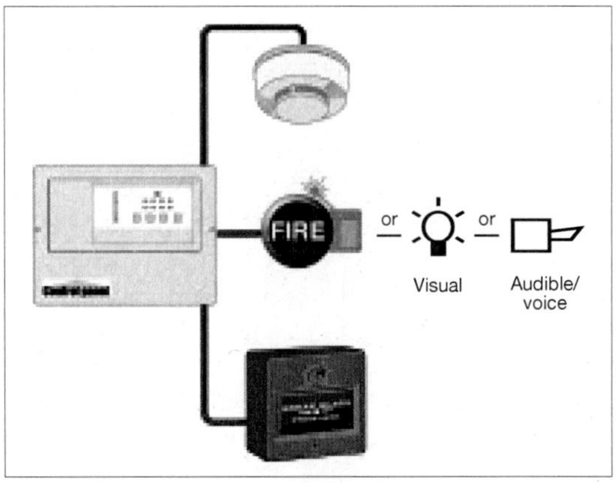

Figure 7: Fire detection and warning system

However, where there are unoccupied areas, or common corridors and circulation spaces in multi-occupied premises, in which a fire could develop to the extent that escape routes could be affected before the fire is discovered, an automatic fire detection system may be necessary.

You may need to consider special arrangements for times when people are working alone, are disabled, or when your normal occupancy patterns are different, e.g. when maintenance staff or other contractors are working at the weekend.

In large or complex premises, particularly those accommodating large numbers of people, such as department stores and multi-storey office blocks, it is likely that a more sophisticated form of warning and evacuation, possibly phased, should be provided.

False alarms from electrical fire warning systems are a major problem (e.g. malicious activation of manual call points) and result in many unwanted calls to the fire and rescue service every year. To help reduce the number of false alarms, the design and location of activation devices should be reviewed against the way the premises are currently used.

If you are not sure whether your current arrangements are adequate, see the additional guidance on fire warning systems in Part 2, Section 2.

Checklist

- Can the existing means of detection ensure a fire is discovered quickly enough for the alarm to be raised in time for all the occupants to escape to a place of total safety? ☐
- Are the detectors of the right type and in the appropriate locations? ☐
- Can the means of warning be clearly heard and understood by everyone throughout the whole building when initiated from a single point? ☐
- Are there provisions for people or locations where the alarm cannot be heard? ☐
- If the fire-detection and warning system is electrically powered, does it have a back-up power supply? ☐

3.4.2 Firefighting equipment and facilities

Firefighting equipment can reduce the risk of a small fire, e.g. a fire in a waste-paper bin, developing into a large one. The safe use of an appropriate fire extinguisher to control a fire in its early stages can also significantly reduce the risk to other people in the premises by allowing people to assist others who are at risk.

This equipment will usually comprise enough portable extinguishers that must be suitable for the risk.

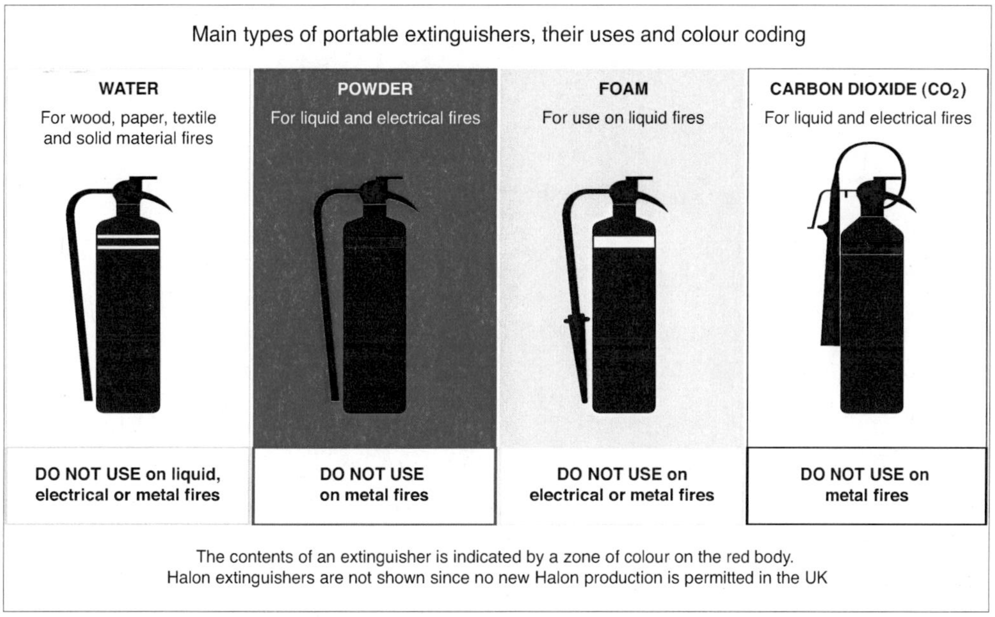

Figure 8: Types of fire extinguishers

In small premises, having one or two portable extinguishers of the appropriate type, readily available for use, may be all that is necessary. In larger, more complex premises, a number of portable extinguishers may be required and they should be sited in suitable locations, e.g. on the escape routes at each floor level. It may also be necessary to indicate the location of extinguishers by suitable signs.

Some premises will also have permanently installed firefighting equipment such as hose reels for use by trained staff or firefighters.

People with no training should not be expected to attempt to extinguish a fire. However, all staff should be familiar with the location and basic operating procedures for the equipment provided, in case they need to use it. If your fire strategy means that certain people, e.g. fire marshals, will be expected to take a more active role, then they should be provided with more comprehensive training.

Other fixed installations and facilities to assist firefighters, such as dry rising mains and access for fire engines, or automatically operated, fixed fire suppression systems such as sprinklers and gas or foam flooding systems may also have been provided.

Where these have been required by law, e.g. the Building Regulations or local Acts, such equipment and facilities must be maintained.

Similarly, if provided for other reasons, e.g. insurance, it is good practice to ensure that they are properly maintained.

In most cases it will be necessary to consult a competent service engineer. Keeping records of the maintenance carried out will help you demonstrate to the enforcing authority that you have complied with fire safety law.

Appendix A.1 provides a sample fire safety maintenance checklist you can use.

For further guidance on portable fire extinguishers see Part 2, Section 3.1, for fixed firefighting installations, Part 2, Section 3.2 and other facilities (including those for firefighters) see Part 2, Section 3.3.

Checklist

- Are the extinguishers suitable for the purpose? ☐
- Are there enough extinguishers sited throughout the premises at appropriate locations? ☐
- Are the right types of extinguishers located close to the fire hazards and can users get to them without exposing themselves to risk? ☐
- Are the extinguishers visible or does their position need indicating? ☐
- Have you taken steps to prevent the misuse of extinguishers? ☐
- Do you regularly check any other equipment provided to help maintain the escape routes? ☐
- Do you carry out daily checks to ensure that there is clear access for fire engines? ☐
- Are those who test and maintain the equipment competent to do so? ☐
- Do you have the necessary procedures in place to maintain any facilities that have been provided for the safety of people in the building (or for the use of firefighters, such as access for fire engines and firefighting lifts)? ☐

3.4.3 Escape routes

Once a fire has started, been detected and a warning given, everyone in your premises should be able to escape to a place of total safety unaided and without the help of the fire and rescue service. However, some people with disabilities and others with special needs may need from staff who will need to be designated for the purpose.

Escape routes should be designed to ensure, as far as possible, that any person confronted by fire anywhere in the building, should be able to turn away from it and escape to a place of reasonable safety, e.g. a protected stairway. From there they will be able to go directly to a place of total safety away from the building.

In offices those who require special assistance (e.g. very young children in a creche and some people with disabilities) could be accommodated on the same level as the final exit from the premises to facilitate escape. Where they need assistance to evacuate, you should make sure that there are sufficient staff to ensure a speedy evacuation.

The level of fire protection that should be given to escape routes will vary depending on the level of risk of fire within the premises and other related factors. Generally, premises that are simple, consisting of a single storey, will require fairly simple measures to protect the escape routes, compared to a large multi-storey building, which would require a more complex and inter-related system of fire precautions.

When determining whether your premises have adequate escape routes, you need to consider a number of factors, including:

- the type and number of people using the premises;
- escape time;
- the age and construction of the premises;
- the number and complexity of escape routes and exits;
- whether lifts can or need to be used;
- the use of phased or delayed alarm evacuation; and
- assisted means of escape/personal evacuation plans (PEEPS).

The type and number of people using the premises

The people present in your premises will sometimes just be employees, but most of the time will be a mixture of employees and members of the public. Employees can reasonably be expected to have an understanding of the layout of the premises, while members of the public will be unlikely to have knowledge of alternative escape routes.

The number and capability of people present will influence your assessment of the escape routes. You must ensure that your existing escape routes are sufficient and capable of safely evacuating all the people likely to use your premises at any time, including events such as sales. If necessary you may need either to increase the capacity of the escape routes or restrict the number of people in the premises.

Escape time

In the event of a fire, it is important to evacuate people as quickly as possible from the premises. Escape routes in a building should be designed so that people can escape quickly enough to ensure they are not placed in any danger from fire. The time available will depend on a number of factors, including how quickly the fire is detected and the alarm raised, the number of escape routes available, the nature of the occupants and the speed of fire growth. For simplicity the travel distances in Part 2, Table 2 on page 68 takes these factors into account. Part 2, Section 4.1 will help you decide the level of risk in your premises for escape purposes.

The age and construction of the premises

Older buildings may comprise different construction materials from newer buildings, and may be in a poorer state of repair. The materials from which your premises are constructed and the quality of building work and state of repair could contribute to the speed with which any fire may spread, and potentially affect the escape routes the occupants will need to use. A fire starting in a building constructed mainly from combustible material will spread faster than one where fire-resisting construction materials have been used.

If you wish to construct internal partitions or walls in your premises, perhaps to create a sales area or to divide up an office area, you should ensure that any new partition or wall does not obstruct any escape routes or fire exits, extend travel distances or reduce the sound levels of the fire alarm system. Any walls that affect the means of escape should be constructed of appropriate material. (Further technical information is provided in Appendix B.)

Depending on the findings of your fire risk assessment, it may be necessary to protect the escape routes against fire and smoke by upgrading the construction of the floors, ceiling and walls to a fire-resisting standard. You should avoid having combustible wall and ceiling linings in your escape routes. For further information see Appendix B. You may need to seek advice from a competent person. Any structural alterations may require building regulation approval.

The number of escape routes and exits

In general there should normally be at least two escape routes from all parts of the premises but a single escape route may be acceptable in some circumstances (e.g. part of your premises accommodating less than 60 people or where travel distances are limited).

Where two escape routes are necessary and to further minimise the risk of people becoming trapped, you should ensure that the escape routes are completely independent of each other. This will prevent a fire affecting more than one escape route at the same time.

When evaluating escape routes, you may need to build in a safety factor by discounting the largest exit from your escape plan, then determine whether the remaining escape routes from a room, floor or building will be sufficient to evacuate all the occupants within a reasonable time. Escape routes that provide escape in a single direction only may need additional fire precautions to be regarded as adequate.

Exit doors on escape routes and final exit doors should normally open in the direction of travel, and be quickly and easily openable without the need for a key. Checks should be made to ensure final exits are wide enough to accommodate the number of people who may use the escape routes they serve.

Management of escape routes

It is essential that escape routes, and the means provided to ensure they are used safely, are managed and maintained to ensure that they remain usable and available at all times when the premises are occupied. Inform staff in training sessions about the escape routes within the premises.

Figure 9: A blocked corridor with incorrect signage

Corridors and stairways that form part of escape routes should be kept clear and hazard free at all times. Items that may be a source of fuel or pose an ignition risk should not normally be located on any corridor or stairway that will be used as an escape route. Further guidance is available in Part 2, Section 4.

Emergency evacuation of persons with mobility impairment

The means of escape you provide must be suitable for the evacuation of everyove likely to be in your premises. This may require additional planning and allocation of staff roles – with appropriate training. Provsions for the emergency evacuation of disabled persons may include:

- stairways;
- evacuation lifts;
- firefighting lifts;
- horizontal evacuation;
- refuges; and
- ramps.

Use of these facilities will need to be linked to effective management arrangements as part of your emergency plan. The plan should not rely on fire and rescue service involvement for it to be effective.

Further guidance on escape routes is given in Part 2, Section 4.

Checklist

- Is your building constructed, particularly in the case of multi-storey buildings, so that, if there is a fire, heat and smoke will not spread uncontrolled through the building to the extent that that people are unable to use the escape routes? ☐
- Are any holes or gaps in walls, ceilings and floors properly sealed, e.g. where services such as ventilation ducts and electrical cables pass through them? ☐
- Can all the occupants escape to a place of total safety in a reasonable time? ☐
- Are the existing escape routes adequate for the numbers and type of people that may need to use them, e.g. staff, members of the public, young children, and disabled people? ☐
- Are the exits in the right place and do the escape routes lead as directly as possible to a place of total safety? ☐
- If there is a fire, could all available exits be affected or will at least one route from any part of the premises remain available? ☐
- Are the escape routes and final exits kept clear at all times? ☐
- Do the doors on escape routes open in the direction of escape? ☐
- Can all final exit doors be opened easily and immediately if there is an emergency? ☐
- Will everybody be able to safely use the escape routes from your premises? ☐
- Are the people who work in the building aware of the importance of maintaining the safety of the escape routes, e.g. by ensuring that fire doors are not wedged open and that combustible materials are not stored within escape routes? ☐
- Are there any particular or unusual issues to consider? ☐

3.4.4 Emergency escape lighting

People in your premises must be able to find their way to a place of total safety if there is a fire by using escape routes that have enough lighting. Where any escape routes are internal and without windows, or your premises are used during periods of darkness, including early darkness on winter days, then some form of backup to the normal escape route lighting (emergency escape lighting) is likely to be required.

In small premises, where the escape routes are simple and straightforward, borrowed lighting, e.g. from street lamps where they illuminate escape routes, may be acceptable. Where borrowed lighting is not available, suitably placed torches may be acceptable for use by trained staff.

In larger, more complex premises it is likely that a more comprehensive system of electrical automatic emergency escape lighting will be needed to illuminate all the escape routes.

Where people have difficulty seeing conventional signs, a 'way-guidance' system may need to be considered.

Further guidance on emergency escape lighting is given in Part 2, Section 5.

Checklist

- Are your premises used during periods of darkness? ☐
- Will there always be sufficient lighting to safely use escape routes? ☐
- Do you have back-up power supplies for your emergency lighting? ☐

3.4.5 Signs and notices

Signs

Signs must be used, where necessary, to help people identify escape routes, find firefighting equipment and emergency fire telephones. These signs are required under the Health and Safety (Safety Signs and Signals) Regulations 1996[5,6] and must comply with the provisions of those Regulations.

A fire risk assessment that determines that no escape signs are required (because, for example, trained staff will always be available to help members of the public to escape routes), is unlikely to be acceptable to an enforcing authority other than in the smallest and simplest of premises where the exits are in regular use and familiar to employees and visitors.

For a sign to comply with these Regulations it must be in pictogram form (see Figure 10). The pictogram can be supplemented by text if this is considered necessary to make the sign more easily understood, but you must not have a safety sign that uses only text.

Where the locations of escape routes and firefighting equipment are readily apparent and the firefighting equipment is visible at all times, then signs are not necessary. In all other situations it is likely that the fire risk assessment will indicate that signs will be necessary.

Figure 10: Typical fire exit sign

Notices

Notices must be used, where necessary, to provide the following:

- instructions on how to use any fire safety equipment;
- the actions to be taken in the event of fire; and
- help for the fire and rescue service (e.g. location of sprinkler valves or electrical cut-off switches).

All signs and notices should be positioned so that they can be easily seen and understood.

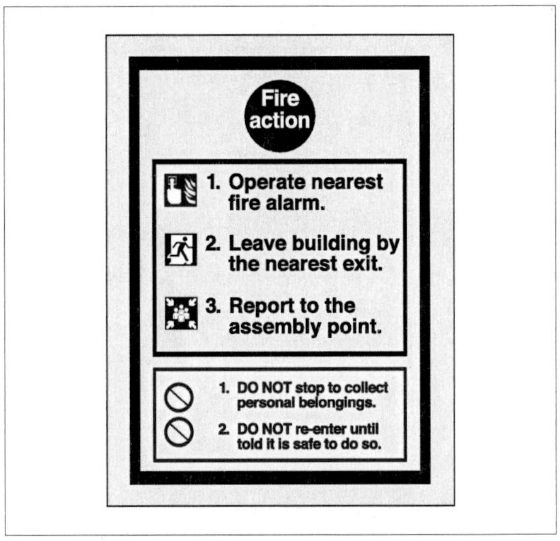

Figure 11: Simple fire action notice

Further guidance on signs and notices is given in Part 2, Section 6.

Checklist

- Where necessary are escape routes and exits, the locations of firefighting equipment and emergency fire telephones indicated by appropriate signs? ☐
- Have you provided notices such as those giving information on how to operate security devices on exit doors, those indicating doors enclosing fire hazards that must be kept shut and fire action notices for staff and other people? ☐
- Are you maintaining all the necessary signs and notices so that they continue to be correct, legible and understood? ☐
- Are you maintaining signs that you have provided for the information of the fire and rescue service, such as those indicating the location of water suppression stop valves and the storage of hazardous substances? ☐

3.4.6 Installation, testing and maintenance

New fire precautions should be installed by a competent person.

You must keep any existing equipment, devices or facilities that are provided in your premises for the safety of people, such as fire alarms, fire extinguishers, lighting, signs, fire exits and fire doors, in effective working order and maintain fire separating elements and the prevention of smoke into escape routes.

You must ensure regular checks, periodic servicing and maintenance are carried out whatever the size of your premises and any defects are put right as quickly as possible.

You, or a person you have nominated, can carry out certain checks and routine maintenance work. Further maintenance may need to be carried out by a competent service engineer. Where contractors are used, third party certification is one method where a reasonable assurance of quality of work and competence can be achieved (see Part 2, Section 8).

The following are examples of checks and tests that should be carried out. The examples of testing and maintenance given are not intended to be prescriptive and other testing regimes may be appropriate.

Daily checks

Remove bolts, padlocks and security devices from fire exits, ensure that doors on escape routes swing freely and close fully and check escape routes to ensure they are clear from obstructions and combustible materials. Check the fire alarm panel to ensure the system is active and fully operational. Where practicable, visually check that emergency lighting units are in good repair and working. Check that all safety signs and notices are legible. (See Appendix B3 for more details on bolts, padlocks and security devices.)

Weekly tests and checks

Test fire-detection and warning systems and manually-operated warning devices weekly following the manufacturer's or installer's instructions. Check the batteries of safety torches and that fire extinguishers and hose reels are correctly located and in apparent working order. Fire pumps and standby diesel engines should be tested for 30 minutes each week.

Monthly tests and checks

Test all emergency lighting systems and safety torches to make sure they have enough charge and illumination according to the manufacturer's or supplier's instructions. This should be at an appropriate time when, following the test, they will not be immediately required.

Check that fire doors are in good working order and closing correctly and that the frames and seals are intact.

Six-monthly tests and checks

A competent person should test and maintain the fire-detection and warning system.

Annual tests and checks

The emergency lighting and all firefighting equipment, fire alarms and other installed systems should be tested and maintained by a competent person.

All structural fire protection and elements of fire compartmentation should be inspected and any remedial action carried out. Specific guidance on the maintenance of timber fire-resisting doors is given in Appendix B2.

Further guidance on maintenance and testing on individual types of equipment and facilities can be found in the relevant Section in Part 2.

Appendix A.1 provides an example of a fire safety maintenance checklist. You will find it of benefit to keep a log book of all maintenance and testing.

Checklist

- Do you regularly check all fire doors and escape routes and associated lighting and signs? ☐
- Do you regularly check all your firefighting equipment? ☐
- Do you regularly check your fire-detection and alarm equipment? ☐
- Are those who test and maintain the equipment competent to do so? ☐
- Do you keep a log book to record tests and maintenance? ☐

Step 3 Checklist

Evaluate, remove, reduce and protect from risks by:

- Evaluating the risk to people in your building if a fire starts ☐
- Removing or reducing the hazards that might cause a fire ☐

  Have you:

  – Removed or reduced sources of ignition? ☐

  – Removed or reduced sources of fuel? ☐

  – Removed or reduced sources of air or oxygen? ☐

Have you removed or reduced the risks to people if a fire occurs by:

– Considering the need for fire detection and for warning? ☐

– Considering the need for firefighting equipment? ☐

– Determining whether your escape routes are adequate? ☐

– Determining whether your lighting and emergency lighting are adequate? ☐

– Checking that you have adequate signs and notices? ☐

– Regularly testing and maintaining safety equipment? ☐

– Considering whether you need any other equipment or facilities? ☐

STEP 4 RECORD, PLAN, INFORM, INSTRUCT AND TRAIN

In Step 4 there are four further elements of the risk assessment you should focus on to address the management of fire safety in your premises. In smaller premises this could be done as part of the day-to-day management, however, as the premises or the organisation get larger it may be necessary for a formal structure and written policy to be developed. Further guidance on managing fire safety is given in Part 2, Section 7 on page 39.

4.1 Record the significant findings and action taken

If you or your organisation employ five or more people, your premises are licensed, or an alterations notice requiring you to do so is in force, you must record the significant findings of your fire risk assessment and the actions you have taken.

Significant findings should include details of:

- The fire hazards you have identified (you don't need to include trivial things like a small tin of solvent based glue).
- The actions you have taken or will take to remove or reduce the chance of a fire occurring (preventive measures).
- Persons who may be at risk, particularly those at greatest risk.
- The actions you have taken or will take to reduce the risk to people from the spread of fire and smoke (protective measures).
- The actions people need to take in case of fire including details of any persons nominated to carry out a particular function (your emergency plan).
- The information, instruction and training you have identified that people need and how it will be given.

You may also wish to record discussions you have had with staff or staff representatives (including trade unions).

Even where you are not required to record the significant findings, it is good practice to do so.

In some very small offices and shops, record keeping may be no more than a few sheets of paper (possibly forming part of a health and safety folder), containing details of significant findings, any action taken and a copy of the emergency plan.

The record could take the form of a simple list which may be supported by a simple plan of the premises (see Figure 12).

In more complex premises, it is best to keep a dedicated record including details of significant findings, any action taken, a copy of the emergency plan, maintenance of fire-protection equipment and training. There is no one 'correct' format specified for this. Further guidance is given in Part 2, Section 7.1.

You must be able to satisfy the enforcing authority, if called upon to do so, that you have carried out a suitable and sufficient fire risk assessment. Keeping records will help you do this and will also form the basis of your subsequent reviews.
If you keep records, you do not need to record all the details, only those that are significant and the action you have taken.

It can be helpful to include a simple line drawing to illustrate your fire precautions (Figure 12). This can also help you check your precautions as part of your ongoing review.

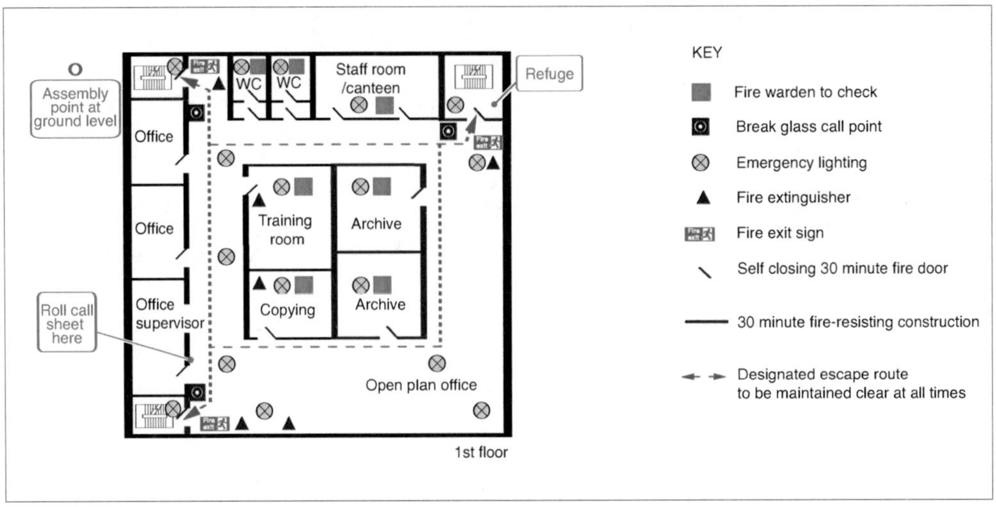

Figure 12: Example of a line drawing showing general fire safety precautions

The findings of your fire risk assessment will help you to develop your emergency plan, the instruction, information and training you need to provide, the co-operation and co-ordination arrangements you may need to have with other responsible people and the arrangements for maintenance and testing of the fire precautions. If you are required to record the significant findings of your fire risk assessment then these arrangements must also be recorded.

Further guidance about fire safety records with an example is given in Part 2, Section 7.1.

Checklist

- Have you recorded the significant findings of your assessment? ☐
- Have you recorded what you have done to remove or reduce the risk? ☐
- Are your records available for inspection by the enforcing authority? ☐

4.2 Emergency plans

You need to have an emergency plan for dealing with any fire situation.

The purpose of an emergency plan is to ensure that the people in your premises know what to do if there is a fire and that the premises can be safely evacuated.

If you or your organisation employ five or more people, or your premises are licensed or an alterations notice requiring it is in force, then details of your emergency plan must be recorded. Even if it is not required, it is good practice to keep a record.

Your emergency plan should be based on the outcome of your fire risk assessment and be available for your employees, their representatives (where appointed) and the enforcing authority.

In small offices and shops the emergency plan may be no more than a fire action notice.

In multi-occupied, larger and more complex offices and shops, the emergency plan will need to be more detailed and compiled only after consultation with other occupiers and other responsible people, e.g. owners, who have control over the building. In most cases this means that a single emergency plan covering the whole building will be necessary. It will help if you can agree on one person to co-ordinate this task.

Further guidance on emergency plans is given in Part 2, Section 7.2.

Checklist

- Do you have an emergency plan and, where necessary, have you recorded the details? ☐
- Does your plan take account of other emergency plans applicable in the building? ☐
- Is the plan readily available for staff to read? ☐
- Is the emergency plan available to the enforcing authority? ☐

4.3 Inform, instruct, co-operate and co-ordinate

You must give clear and relevant information and appropriate instructions to your staff and the employers of other people working in your premises, such as contractors, about how to prevent fires and what they should do if there is a fire.

Any other relevant persons should be given information about the fire safety arrangements as soon as possible.

If you intend to employ a child, you must inform the parents of the significant risks you have identified and the precautions you have taken. You must also co-operate and co-ordinate with other responsible people who use any part of the premises. It is unlikely that your emergency plan will work without this.

Information and instruction

All staff should be given information and instruction as soon as possible after they are appointed and regularly after that. Make sure you include staff who work outside normal working hours, such as contract cleaners or maintenance staff.

The information and instructions you give must be in a form that can be used and understood. They should take account of those with disabilities such as hearing or sight impairment, those with learning difficulties and those who do not use English as their first language.

The information and instruction you give should be based on your emergency plan and must include:

- the significant findings from your fire risk assessment;
- the measures that you have put in place to reduce the risk;
- what staff should do if there is a fire;
- the identity of people you have nominated with responsibilities for fire safety; and
- any special arrangements for serious and imminent danger to persons from fire.

In small premises, where no significant risks have been identified and there are limited numbers of staff, information and instruction may simply involve an explanation of the fire procedures and how they are to be applied. This should include showing staff the fire-protection arrangements, including the designated escape routes, the location and operation of the fire-warning system and any other fire-safety equipment provided, such as fire extinguishers. Fire action notices can complement this information and, where used, should be posted in prominent locations.

In larger premises, particularly those in multi-occupied buildings, you should ensure that written instructions are given to people who have been nominated to carry out a designated safety task, such as calling the fire and rescue service or checking that exit doors are available for use at the start of each working day.

Further guidance on information and instruction to staff, and on working with dangerous substances is given in Part 2, Section 7.3.

Co-operation and co-ordination

In premises that are not multi-occupied you are likely to be solely responsible. However, in buildings owned by someone else, or where there is more than one occupier, and others are responsible for different parts of the building, it is important that you liaise with them and inform them of any significant risks that you have identified. By liaising you can co-ordinate your resources to ensure that your actions and working practices do not place others at risk if there is a fire, and a co-ordinated emergency plan operates effectively.

Where two or more responsible persons share premises in which an explosive atmosphere may occur, the responsible person with overall responsibility for the premises must co-ordinate any measures necessary to protect everyone from any risk that may arise. Employees also have a responsibility to co-operate with their employer so far as it is necessary to help the employer comply with any legal duty.

Further information on co-operation and co-ordination is given in Part 2, Section 7.3.

Checklist

- Have you told your staff about the emergency plan? ☐
- Have you informed guests and visitors about what to do in an emergency? ☐
- Have you identified people you have nominated to do a particular task? ☐
- Have you given staff information about any dangerous substances? ☐
- Do you have arrangements for informing temporary or agency staff? ☐
- Do you have arrangements for informing other employers whose staff are guest workers in your premises, such as maintenance contractors and cleaners? ☐
- Have you co-ordinated your fire safety arrangements with other responsible people in the building? ☐
- Have you recorded details of any information or instructions you have given and the details of any arrangements for co-operation and co-ordination with others? ☐

4.4 Fire safety training

You must provide adequate fire safety training for your staff. The type of training should be based on the particular features of your premises and should:

- take account of the findings of the fire risk assessment;
- explain your emergency procedures;
- take account of the work activity and explain the duties and responsibilities of staff;
- take place during normal working hours and be repeated periodically where appropriate;
- be easily understandable by your staff and other people who may be present; and
- be tested by fire drills.

In small premises this may be no more than showing new staff the fire exits and giving basic training on what to do if there is a fire. In larger premises, such as a supermarket with a high staff turnover and many shift patterns, the organisation of fire safety training will need to be planned.

Your staff training should include the following:

- what to do on discovering a fire;
- how to raise the alarm and what happens then;
- what to do upon hearing the fire alarm;
- the procedures for alerting members of the public and visitors including, where appropriate, directing them to exits;
- the arrangements for calling the fire and rescue service;
- the evacuation procedures for everyone in your office or shop to reach an assembly point at a place of total safety;
- the location and, when appropriate, the use of firefighting equipment;
- the location of escape routes, especially those not in regular use;
- how to open all emergency exit doors;
- the importance of keeping fire doors closed to prevent the spread of fire, heat and smoke;
- where appropriate, how to stop machines and processes and isolate power supplies in the event of a fire;
- the reason for not using lifts (except those specifically installed or nominated, following a suitable fire risk assessment, for the evacuation of people with a disability);
- the safe use of and risks from storing or working with highly flammable and explosive substances; and
- the importance of general fire safety, which includes good housekeeping.

All the staff identified in your emergency plan that have a supervisory role if there is a fire (e.g. heads of department, fire marshals or wardens and, in larger offices and shops, fire parties or teams), should be given details of your fire risk assessment and receive additional training.

Further guidance on training and how to carry out a fire drill is given in Part 2, Section 7.4.

Checklist

- Have your staff received fire safety training? ☐
- Have you carried out a fire drill recently? ☐
- Are employees aware of specific tasks if there is a fire? ☐
- Are you maintaining a record of training sessions? ☐
- Do you carry out joint training and fire drills in multi-occupied buildings? ☐
- If you use or store hazardous or explosive substances have your staff received appropriate training? ☐

STEP 5 REVIEW

You should constantly monitor what you are doing to implement the fire risk assessment to assess how effectively the risk is being controlled.

If you have any reason to suspect that your fire risk assessment is no longer valid or there has been a significant change in your premises that has affected your fire precautions, you will need to review your assessment and if necessary revise it. Reasons for review could include:

- changes to work processes or the way that you organise them, including the introduction of new equipment;
- alterations to the building, including the internal layout;
- substantial changes to furniture and fixings;
- the introduction, change of use or increase in the storage of hazardous substances;
- the failure of fire precautions, e.g. fire-detection systems and alarm systems, life safety sprinklers or ventilation systems;
- significant changes to displays or quantities of stock;
- a significant increase in the number of people present; and
- the presence of people with some form of disability.

You should consider the potential risk of any significant change before it is introduced. It is usually more effective to minimise a risk by, for example, ensuring adequate, appropriate storage space for an item before introducing it to your premises.

Do not amend your assessment for every trivial change, but if a change introduces new hazards you should consider them and, if significant, do whatever you need to do to keep the risks under control. In any case you should keep your assessment under review to make sure that the precautions are still working effectively. You may want to re-examine the fire prevention and protection measures at the same time as your health and safety assessment.

If a fire or 'near miss' occurs, this could indicate that your existing assessment may be inadequate and you should carry out a re-assessment. It is good practice to identify the cause of any incident and then review and, if necessary, revise your fire risk assessment in the light of this.

Records of testing, maintenance and training etc. are useful aids in a review process. See Appendix A.1 for an example.

Alterations notices

If you have been served with an 'alterations notice' check it to see whether you need to notify the enforcing authority about any changes you propose to make as a result of your review. If these changes include building work, you should also consult a building control body.

END OF PART 1

You should now have completed the five-step fire risk assessment process, using the additional information in Part 2 where necessary. In any review you may need to revisit Steps 1 to 4.

Part 2 Further guidance on fire risk assessment and fire precautions

Managing fire safety

Good management of fire safety in your premises is essential to ensure that any fire safety matters that arise are always effectively addressed. In small premises this can be achieved by the manager or owner responsible for maintaining and planning fire safety in conjunction with general health and safety.

In larger premises, it is good practice for a senior manager to have overall responsibility for fire safety. It may be appropriate for this responsibility to be placed with the person designated with overall responsibility for health and safety.

An organisation's fire safety policy should be flexible enough to allow modification. It should be recognised that fire safety operates at all levels within an organisation and therefore those responsible for fire safety should be able to develop, where necessary, a local action plan for their premises.

The organisation's policy should be set out in writing and may cover such things as:

- who will hold the responsibility for fire safety at board level;
- who will be the responsible person for each of their premises (this will be the person who has overall control usually the manager);
- the arrangement whereby those responsible for fire safety will, where necessary, nominate in writing specific people to carry out particular tasks if there is a fire; and
- arrangements to monitor and check that individual persons responsible for fire safety are meeting the requirements of the fire safety law.

You should have a plan of action to bring together all the features you have evaluated and noted from your fire risk assessment so that you can logically plan what needs to be done. It should not be confused with the emergency plan, which is a statement of what you will do if there is a fire.

The plan of action should include what you intend to do to reduce the hazards and risks you have identified and to implement the necessary protection measures.

You will need to prioritise these actions to ensure that any findings which identify people in immediate danger are dealt with straight away, e.g. unlocking fire exits. In other cases where people are not in immediate danger but action is still necessary, it may be acceptable to plan this over a period of time.

Before admitting the public to your premises you need to ensure that all of your fire safety provisions are in place and in working order, or, if not, that alternative arrangements are in place. Constant checks are needed while the public are present, and again after they have left. Detailed recommendations are given in BS 5588-12.[52]

The guidance in Part 2 provides additional information to:

- ensure good fire safety management by helping you establish your fire prevention measures, fire precautions and fire safety procedures (systems equipment and plans); and
- assist you to carry out your fire safety risk assessment and identify any issues that need attention.

Section 1 Further guidance on fire risks and preventative measures

This section provides further information on evaluating the risk of a fire and its prevention in your premises. You should spend time developing long-term workable and effective strategies to reduce hazards and the risk of a fire starting. At its simplest this means separating flammable materials from ignition sources.

You should consider:

- housekeeping;
- storage;
- dangerous substances storage, display and use;
- equipment and machinery;
- electrical safety;
- smoking;
- managing building work and alterations;
- existing layout and construction;
- particular hazards in corridors and stairways used as escape routes;
- insulated core panels;
- restricting the spread of fire and smoke;
- fire-resisting structures;
- arson; and
- help for people with special needs.

1.1 Housekeeping

Good housekeeping will lower the chances of a fire starting, so the accumulation of combustible materials in all premises should be monitored carefully. Good housekeeping is essential to reduce the chances of escape routes and fire doors being blocked or obstructed.

Keep waste material in suitable containers before it is removed from the premises. If bins, particularly wheeled bins, are used outside, secure them in a compound to prevent them being moved to a position next to the building and set on fire. Never place skips against a building (Figure 13) – they should normally be a minimum of 6m away from any part of the premises.

If you generate a considerable quantity of combustible waste material then you may need to develop a formal plan to manage this effectively.

In higher risk areas you need to make sure arrangements are in place for close down, e.g. checking all appliances are turned off and combustible waste has been removed.

Figure 13: Bins under stairway (courtesy of Cheshire fire and rescue service)

1.2 Storage

Many of the materials found in your premises will be combustible. If your premises have inadequate or poorly managed storage areas then the risk of fire is likely to be increased (Figure 14). The more combustible materials you store the greater the source of fuel for a fire. Poorly arranged storage could prevent equipment such as sprinklers working effectively.

Combustible materials are not just those generally regarded as highly combustible, such as polystyrene, but all materials that will readily catch fire. However, by carefully considering the type of material, the quantities kept and the storage arrangements, the risks can be significantly reduced.

Figure 14: An example of poor storage

In offices, the retention of large quantities of paper records, especially if not filed away in proprietary cabinets, can increase the fire hazard. Such readily available flammable material makes the potential effect of arson more serious.

> Case study
>
> A national retailer with many units in covered shopping centres made large centrally controlled stock deliveries to those units. The arrival of such quantities of stock, while economically efficient, left local unit managers with inadequate storage space and often led to an increased fire risk as boxes were piled up in unsuitable areas. The need for additional storage space or smaller, more frequent deliveries was clear and should have been anticipated.

Many shops will take great care to present an efficient and attractive image in the retail area, while other areas are neglected and allowed to become over-stocked or dumping areas for unsold material.

To reduce the risk, store excess materials and stock in a dedicated storage area, storeroom or cupboard. Do not store excess stock in areas where the public would normally have access.

Do not pile combustible material against electrical equipment or heaters, even if turned off for the summer, and do not allow smoking in areas where combustible materials are stored.

Consider how stock is displayed in shops and evaluate any additional risk of fire that it generates. For example, rugs stacked on the floor on top of each other would not present a high fire risk, but rolls of carpet stored vertically up against a wall or hung on displays present a vertical surface for fire to spread rapidly upwards. The display of large quantities of clothing on vertical hangers is also likely to increase the risk of rapid fire development.

Your fire risk assessment should also consider any additional risk generated by seasonal products such as fireworks and Christmas decorations.

Consider the following to reduce these risks:

- ensure storage and display areas are adequately controlled and monitored;
- use fire-resistant display materials wherever possible (suppliers should be able to provide evidence of this); and
- ensure electrical lighting used as part of the display does not become a potential source of ignition.

Voids

Voids (including roof voids) should not be used for the storage of combustible material. Such voids should be sealed off or kept entirely open to allow for easy access for inspection and the removal of combustible materials.

Combustible waste and packaging

Delivery of some goods results in large quantities of combustible waste and packaging. The sighting use and removal of these materials needs to be carefully managed to ensure that they cannot come into contact with potential ignition sources and to not cause obstructions.

1.3 Dangerous substances: storage, display and use

Specific precautions are required when handling and storing dangerous substances to minimise the possibility of an incident. Your supplier should be able to provide detailed advice on safe storage and handling, however, the following principles will help you reduce the risk from fire:

- substitute highly flammable substances and materials with less flammable ones;
- reduce the quantity of dangerous substances to the smallest reasonable amount necessary for running the business or organisation;

- correctly store dangerous substances, e.g. in a fire-resisting enclosure. All flammable liquids and gases should ideally be locked away, especially when the premises are unoccupied, to reduce the chance of them being used in an arson attack; and
- ensure that you and your employees are aware of the fire risk the dangerous substances present and the precautions necessary to avoid danger.

Additional general fire precautions may be needed to take account of the additional risks that may be posed by the storage and use of these substances.

Certain substances and materials are by their nature, highly flammable, oxidising or potentially explosive. These substances are controlled by other legislation in addition to fire safety law, in particular the Dangerous Substances and Explosive Atmospheres Regulations2002[7] (also see the HSE's *Approved Code of Practice and guidance*[8]).

Flammable liquids

Highly flammable liquids present a particularly high fire risk. For example, a leak from a container of flammable solvents, such as methylated spirit, will produce large quantities of heavier-than-air flammable vapours. These can travel large distances, increasing the likelihood of their reaching a source of ignition well away from the original leak, such as a basement containing heating plant and/or electrical equipment on automatic timers.

The risk is reduced by ensuring the storage and use of highly flammable liquids is carefully managed, that materials contaminated with solvent are properly disposed of (Figure 15) and when not in use, they are safely stored. Up to 50 litres may be stored in a fire-resisting cabinet or bin that will contain any leaks (Figure 16).

In retail premises the quantity of flammable liquids on display should be kept to the minimum to meet business needs.

Figure 15: A fire-resisting pedal bin for rags

Figure 16: A 50 litre storage bin for flammables

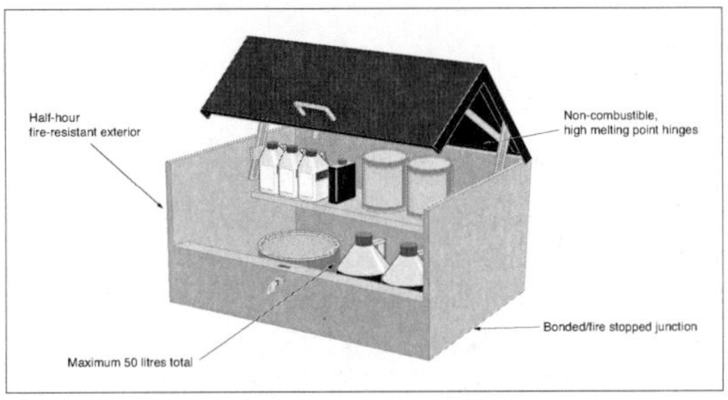

There should be no potential ignition sources in areas where flammable liquids are used or stored and flammable concentrations of vapours may be present. Any electrical equipment used in these areas, including fire alarm and emergency lighting systems, needs to be suitable for use in flammable atmospheres. In such situations, it is recommended that you should seek advice from a competent person.

> Case study
>
> There have been significant fires in retail premises in displays of white spirit and similar products. The plastic containers readily melt in the fire, spilling their contents, fuelling very rapid fire growth. To minimise this risk, you need to ensure that the stock is kept to the minimum necessary for business needs, especially in areas to which the public have access. Storage areas should be carefully chosen, so that they are away from sources of ignition and heating in the building. This will also enable your staff to readily see any incidents, such as spillages and leaks. Your staff should be trained in what do in an emergency and the necessary equipment and arrangements need to be put in place to enable them to do this safely.

LPG storage

Where LPG in cylinders or cartridges is present, you need to take particular care to minimise the possibility of its involvement in a fire. The total amount of LPG for display or demonstration in retail areas should be kept to the minimum necessary to meet business needs. In no circumstances should it exceed 70kg and this should be reduced to 15kg if the retail premises are under residential accommodation or part of a multi-use building, unless the two are separated by a substantial partition that is imperforate and provides at least 60 minutes fire resistance.

Locate cylinders and cartridges in a safe and secure place where they:

- cannot be interfered with;
- can be kept upright (with valve protection fitted);
- are away from sources of ignition and/or readily ignitable materials;
- are away from any corrosive, toxic or oxidant materials; and;
- are away from stairways, exit doors and places where they may obstruct the means of escape.

Quantities of LPG in excess of the amounts indicated for retail areas should be stored in a properly designed and located storage area that is dedicated for LPG.

Bulk storage tanks for LPG should be designed, installed and located in accordance with industry guidance.

Advice on the use of LPG for heating is given in Section 1.4.

Further guidance on the safe storage of LPG is available from your supplier or the Liquefied Petroleum Gas Association's Code of Practice.[9]

Fireworks

Fireworks can cause fires and explosions in shops. They have the potential for a violent release of pressure and heat that can cause severe harm to people and damage to a building.

Fireworks can be ignited or detonated by contact with ignition sources or by contamination, where other chemicals or water cause the material to become unstable.

If you sell fireworks you have certain obligations to the public and your staff. You must:

- be 'registered' or obtain a licence to store fireworks;
- obtain a 'licence to sell fireworks' if you intend to sell outside the usual fireworks period;
- store fireworks safely;
- not sell fireworks to under 18s; and
- review your risk assessment and, where necessary, take additional measures.

Preventative measures include:

- store and display fireworks in appropriate containers, display cabinets;
- keep all fireworks in a dry place;
- do not store an excessive amount of fireworks either in the shop or stockroom and do not store more than your licence permits;
- do not store the fireworks near other articles that could spread the fire;
- do not store fireworks near an escape route;
- exclude sources of ignition and in particular, stop people smoking anywhere near the fireworks; and
- avoid all unnecessary handling of fireworks;

Further guidance on the storage and sale of fireworks is available on the DTI website www.dti.gov.uk/fireworks.

Aerosols

Some aerosols can contain flammable products stored at pressure and they can present a high level of hazard. When ignited they can explode, produce fireballs and rocket to distances of 40m. Their presence in premises can make it unsafe for firefighters to enter a building and they have the potential for starting multiple fires.

The following should be considered to reduce these risks:

- All staff involved in the movement, storage and display of aerosol cans should be adequately instructed, trained and supervised.
- Damaged and leaking aerosol cans should be removed immediately to a safe, secure, well ventilated place prior to disposal. Powered vehicles should not be used to move damaged stock, unless specially adapted for use in flammable atmospheres. Arrangements should be made for disposal at a licenced waste management facility.

1.4 Equipment and machinery

Common causes of fire in equipment are:

- allowing ventilation points to become clogged or blocked, causing overheating;
- inadequate cleaning of heat-shrink packaging equipment, such as that used in in-store bakeries;
- allowing extraction equipment in catering environments to build up excessive grease deposits;
- misuse or lack of maintenance of cooking equipment and appliances; and
- disabling or interfering with automatic or manual safety features and cut-outs.

All machinery, apparatus and office equipment should be properly maintained by a competent person. Appropriate signs and instructions on safe use may be necessary.

Heating

Individual heating appliances require particular care if they are to be used safely, particularly those which are kept for emergency use during a power cut or as supplementary heating during severe weather. The greatest risks arise from lack of maintenance and staff unfamiliarity with them. Heaters should preferably be secured in position when in use and fitted with a fire guard if appropriate.

As a general rule, convector or fan heaters should be preferred to radiant heaters because they present a lower risk of fire and injury. The following rules should be observed:

- All heaters should be kept well clear of combustible materials and where they do not cause an obstruction.
- Heaters which burn a fuel should be sited away from draughts.
- Portable fuel burning heaters (including bottled gas (LPG)) should only be used in exceptional circumstances and if shown to be acceptable in your risk assessment.

All gas heating appliances should be used only in accordance with manufacturer's instructions and should be serviced annually by a competent person.

In general, staff should be discouraged from bringing in their own portable heaters and other electrical equipment (e.g. kettles) into the premises.

Cooking processes

Typical installations used in cooking processes include deep fat fryers, ovens, grills, surface cookers, ductwork, flues, filters, hoods, extract and ventilation ducts and dampers.

These cooking processes can operate with high temperatures, involving large quantities of oil and combustible food stuffs. Heat sources used for cooking processes include: gas, electric and microwave. The main cause of fire are ignition of cooking oil, combustion of crumbs and sediment deposits, and ductwork fires from a build up of fats and grease.

The siting of cooking processes close to insulated core panels with combustible insulation (see Section 1.10) can lead to the likely ignition of the panels and consequent rapid fire spread to other parts of the building. This practice should therefore be avoided.

The following should be considered to reduce the risk from cooking processes:

- regular cleaning to prevent build-up of crumbs and other combustible material;
- fire resisting containers for waste products;
- a fire suppression system capable of controlling an outbreak of fire;
- monitored heat/oil levels, even after the cooking process is complete and installation of temperature control/cut-off/shut off devices as appropriate;
- duct, joints and supports able to withstand high cooking temperatures;
- seperation from wall and ceiling panels (with combustible insulation) e.g. 2.5m for walls, 4m for ceilings;
- insulation of ducts to prevent heating/ignition of nearby combustible wall and ceiling materials;
- a regular programme for inspection and cleaning;
- a programme of electrical and mechanical maintenance; and
- annual service of all gas heating appliances by a competent person.

1.5 Electrical safety

Electrical equipment can be a significant cause of accidental fires in shops and offices. The main causes are:

- overheating cables and equipment, e.g. due to overloading circuits, bunched or coiled cables or impaired cooling fans;
- incorrect installation or use of equipment;
- little or no maintenance and testing of equipment;
- incorrect fuse ratings;
- damaged or inadequate insulation on cables or wiring;
- combustible materials being placed too close to electrical equipment which may give off heat even when operating normally or may become hot due to a fault;
- arcing or sparking by electrical equipment; and
- embrittlement and cracking of cable sheathing in cold environments.

All electrical equipment should be installed and maintained in a safe manner by a competent person. If portable electrical equipment is used, including items brought into a workplace by staff, then your fire risk assessment should ensure that it visually inspected and undergoes portable appliance testing ('PAT') at intervals suitable for the type of equipment and its frequency of use (refer to HSE guidance[10]). If you have any doubt about the safety of your electrical installation then you should consult a competent electrician.

Issues to consider include:

- overloading of equipment;
- correct fuse ratings;
- PAT testing and testing of fixed installations;
- protection against overloading of installation;
- protection against short circuit;
- insulation, earthing and electrical isolation requirements;
- frequency of electrical inspection and test;
- temperature rating and mechanical strength of flexible cables;
- portable electrical equipment;
- physical environment in which the equipment is used (e.g. wet or dusty atmospheres); and
- suitable use and maintenance of personal protective equipment.

All electrical installations should be regularly inspected by a competent electrical engineer appointed by you, or on your behalf, in accordance with the Electricity at Work Regulations 1989 (EAW Regulations).[53] The use of low voltage equipment should conform to the requirements of the Electrical Equipment (Safety) Regulations 1994,[54] including the requirement to be CE marked.

1.6 Smoking

Carelessly discarded cigarettes and other smoking materials are a major cause of fire. A cigarette can smoulder for several hours, especially when surrounded by combustible material. Many fires are started several hours after the smoking materials have been emptied into waste bags and left for future disposal.

Consider prohibiting smoking in your premises other than in the designated smoking areas. Display suitable signs throughout the premises informing people of the smoking policy and the locations where smoking is permitted.

In those areas where smoking is permitted, provide non-combustible deep and substantial ashtrays to help prevent unsuitable containers being used. Empty all ashtrays daily into a metal waste bin and take it outside. It is dangerous to empty ashtrays into plastic waste sacks which are then left inside for disposal later.

1.7 Managing building work and alterations

Fires are more frequent when buildings are undergoing refurbishment or alteration.

You should ensure that, before any building work starts, you have reviewed the fire risk assessment and considered what additional dangers are likely to be introduced. You will need to evaluate the additional risks to people, particularly in those buildings that continue to be occupied. Lack of pre-planning can lead to haphazard co-ordination of fire safety measures.

You should liaise and exchange information with contractors who will also have a duty under the Construction (Health, Safety and Welfare) Regulations 1996[11, 12] to carry out a risk assessment and inform you of their significant findings and the preventive measures they may employ. This may be supported by the contractors' agreed work method statement. The designer should also have considered fire safety as part of the Construction (Design and Management) Regulations 1994 (the CDM Regulations).[55]

You should continuously monitor the impact of the building work on the general fire safety precautions, such as the increased risk from quantities of combustible materials and accumulated waste and maintaining adequate means of escape. You should only allow the minimum materials necessary for the work in hand within or adjacent to your building.

Additional risks can include:

- hot work such as flame cutting, welding, soldering, or paint stripping;
- temporary electrical equipment;
- blocking of escape routes, including external escape routes;
- introduction of combustibles into an escape route;
- loss of normal storage facilities;
- fire safety equipment, such as automatic fire-detection systems becoming affected;
- fire-resisting partitions being breached or fire doors being wedged open (see Appendix B1 for information on fire-resisting separation); and
- additional personnel who may be unfamiliar with the premises.

Activities such as welding, flame cutting, use of blow lamps or portable grinding equipment can pose a serious fire hazard and need to be strictly controlled when carried out in areas near flammable materials. This can be done by having a written permit to work for the people involved (whether they are your employees or those of the contractor).[12]

A permit to work is appropriate in situations of high hazard/risk and, for example, where there is a need to:

- ensure that there is a formal check confirming that a safe system of work is being followed;
- co-ordinate with other people or activities;
- provide time-limits when it is safe to carry out the work; and
- provide specialised personal protective equipment (such as breathing apparatus) or methods of communication.

You must notify the fire and rescue service about alterations in your premises if an alterations notice is in force.

Further guidance on fire safety during construction work is available from the HSE[56,57] and the Fire Protection Association.[58]

1.8 Existing layout and construction

In many shops, and increasingly in offices, the design is for open-plan areas allowing customers and employees to move freely throughout the floor.

Traditionally, occupants are advised to shut doors when escaping from a fire but in open-plan areas there are few doors to shut. In these areas the fire, and especially the smoke, may spread faster than expected.

To assess the risk in your premises you need to evaluate the construction and layout of your premises. This does not mean a structural survey, unless you suspect that the structure is damaged or any structural fire protection is missing or damaged, but rather an informed look around to see if there are any easy paths through which smoke and fire may spread and what you can do to stop that. In general, older buildings will have more void areas, possibly hidden from view, which will allow smoke and fire to spread away from its source. Whatever your type of building, you may need to consider typical situations that may assist the spread of fire and smoke such as:

- vertical shafts, e.g. lifts, open stairways, dumb waiters or holes for moving stock around;
- false ceilings, especially if they are not fire-stopped above walls;
- voids behind wall panelling;
- unsealed holes in walls and ceilings where pipe work, cables or other services have been installed; and
- doors, particularly to stairways, which are ill-fitting or routinely left open.

1.9 Particular hazards in corridors and stairways used as escape routes

Items that are a source of fuel, pose an ignition risk, or are combustible and likely to increase the fire loading or spread of fire, should not be located on any corridor or, stairway or circulation space that will be used as an escape route. Such items include:

- portable heaters, e.g. bottled gas (LPG) or electric radiant heaters and electric convectors or boilers;
- gas cylinders for supplying heaters;
- cooking appliances; and
- unenclosed gas pipes, meters, and other fittings.

However, where more than one escape route is available and depending on the findings of your risk assessment, items such as those below may be acceptable if the minimum exit widths are maintained and the item presents a relatively low fire risk:

- non-combustible lockers;
- vending machines;
- small items of electrical equipment (e.g. photocopiers); and
- small coat racks and/or small quantities of upholstered furniture which meets BS 7176 or the Furniture and Furnishings (Fire) (Safety) Regulations 1988.

1.10 Insulated core panels

Many buildings have insulated core panels as exterior cladding or for internal structures (Figure 17) and partitions. The food industry, in particular, uses insulated core panels because they are easy to clean and facilitate consistent temperature control within the premises. The simple construction of these panels enables alterations and for additional internal partitions to be erected with minimum disruption to business.

Figure 17: Insulated core panels – internal

They normally consist of a central insulated core, sandwiched between an inner and outer metal skin. There is no air gap. The external surface is then normally coated with a PVC covering to improve weather resistance or the aesthetic appeal of the panel. The central core can be made of various insulating materials, ranging from virtually non-combustible through to highly combustible. Differing fire hazards are associated with common types of insulation, when the panels are subjected to certain temperatures. Typical examples are:

- Mineral rock/modified phenolic will produce surface char and little smoke or gaseous combustion products, at temperatures above 230°C.
- Polyisocyanurate (PIR)/polyurethane (PUR) will char and will generate smoke and gaseous combustion products, at temperatures above 430°C PIR and 300°C PUR.
- Expanded polystyrene (EPS) will melt and will generate smoke and gaseous combustion products, at temperatures above 430°C PIR.

Insulation charring can lead to panel delamination/ collapse, and the gaseous combustion products can fill areas with the toxic gases carbon monoxide and styrene.

A number of fires in buildings where insulated panels have been used extensively in the fabric of the building have highlighted the particular dangers that may be associated with this form of construction, i.e. where the fabric of the building can contribute to the fire hazards.

In a fire the following may occur:

- early buckling and falling away of the facing materials;
- burning of the combustible insulating material;
- production of large quantities of dense, toxic smoke;
- rapid heat generation;
- early loss of structural strength can result if the system has not been properly designed, and this can lead to the collapse of the wall, partition or ceiling;
- there may be cavities in older buildings where the panels are used as an internal envelope, enabling fire to spread unnoticed and possibly unchecked by fire barriers.

Once installed it is difficult to identify the core material of a panel and its potential fire hazard.

The following best practise can help you reduce risks associated with insulated panels.

- Do not install heating appliances, such as ovens, against the panels. Operate a clear distance policy for cooking systems.
- Control ignition sources that are adjacent to, or penetrating the panels.
- Control hot working.
- Check for damage to heater tapes used to prevent ice build-up at doors.
- Do not store highly combustible materials against panels or allow rubbish to collect against panels.
- Have damaged panels or sealed joints repaired immediately and make sure that jointing compounds or gaskets used around the edges of the panels are in good order.
- Check where openings have been made for doors, windows, cables and ducts to ensure that these have been effectively sealed and the inner core has not been exposed.
- Check that there has been no mechanical damage and repair any that has occured, e.g. caused by mobile equipment such as fork lift trucks.
- Ensure that any loads, such as storage and equipment, are only supported by panels which have been designed and installed to perform this function.
- Check that the inner and outer skins are adhering tightly to the core.
- Ensure that the panels are correctly secured to the structure or are designed to be independently structurally secure. One solution is to ensure the retaining brackets bolt the panel to a support frame through the outer and inner skins;
- Ensure large roof cavities are appropriately protected, e.g. escape routes are clear, signed and have limited travel distance, and fire warning systems are audible.

The use of combustible panels in areas of buildings with a high life risk, e.g. where large numbers of people are present, should be carefully considered. Your fire risk assessment may need to be revised to ensure that any increased risk resulting from this type of construction is considered.

The potential for fire development involving mineral fibre cores is less than that for panels containing polymeric cores. Therefore, in areas where there is considerable life risk, it may be appropriate to consider replacing combustible panels, providing a fire suppression system or installing non-combustible fire breaks within or between the panels at suitable intervals.

Insulated core panels should be installed by a competent person in accordance with industry guidance.

Guidance on the design, construction, specification and fire management of insulated core panels has been published by the International Association of Cold Storage Contractors.[59]

1.11 Restricting the spread of fire and smoke

To reduce the risk to people if there is a fire, you need to consider how to control or restrict the spread of fire and smoke. The majority of people who die in fires are overcome by the smoke and gases.

It is important ensure that, in the event of fire, the rate of fire growth is restricted in its early stages. It should also be noted that most measures which restrict the rate of fire growth in its early stages will also serve to restrict the fire spread in its later stages.

Case study

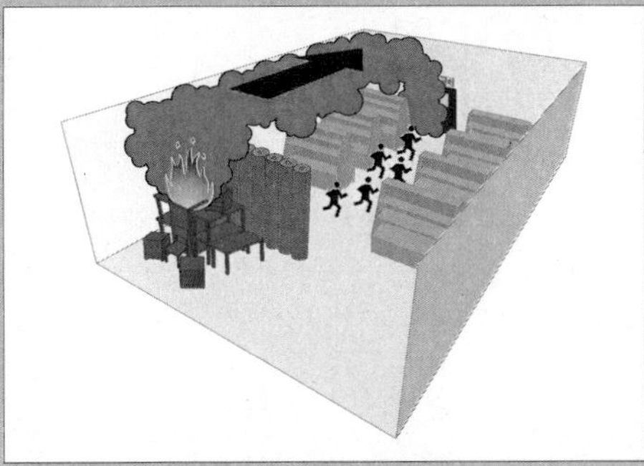

How the goods and fixtures in offices and shops are stored and used is important in assessing the risk to people. This case study is an example of an actual fire which resulted in significant loss of life. It shows rolls of carpet being used to form a barrier behind which furniture and stock is being stored. A fire started and the occupants were not aware of it as it was out of sight. The fire had a good vertical surface of readily combustible material to move quickly upwards. By the time people became aware of the fire the smoke had already moved to the far end of the shop and descended, blocking off the escape route. Alternative storage arrangements may have prevented this fire.

Display materials and decorations

Displays are often located in corridors, entrance foyers and shop floors, and generally comprise materials such as paper, cardboard and plastic which provide a means for the rapid spread of fire. You should assess how these materials could contribute to the development of a fire in the area where they are located. To reduce the risk of fire spread, you should consider the following:

- avoid the use of displays in corridors and foyers;
- minimise the size and number of display areas to discrete, separated areas;
- treating displays with proprietary flame-retardant sprays;
- the use of display boxes;
- keep displays away from curtains, light fittings and heaters;
- keep displays away from ceiling voids which may lack fire barriers; and
- ensure that there are no igntion sources in the vicinity.

Staff information should be confined to appropriately located display boards in areas away from escape routes. Display boards may be used on escape routes as long as they are no bigger than $1m^2$ or have been enclosed in a sealed display case.

Catering facilities

Wherever possible any extensive catering facilities, particularly those with deep fryers, should be located in a seperate building, or alternatively, seperated from the remainder of the building by fire doors and fire resisting construction and provided with adequate ventilation.

Where flues pass through any part of the structure, the structure should be protected by fire resisting construction. Where fire shutters are used these should be capable of operating both manually and by fusible link. Where a fire detection and warning system is installed, the fire shutter should also be designed to close on activation of the system via a controlled geared mechanism.

Fire-resisting structures

Many buildings are divided into different areas by fire doors and fire-resisting walls and floors. These are partly designed to keep a fire within one area, giving people more time to escape. You will need to identify which doors, walls and floors in your building are fire-resisting. There may be information available from when the building was built, if alterations have been made, or from a previously held fire certificate.

High-risk areas should be seperated from the rest of the premises by 30-minute fire-resisting construction.

Normally if there are fire doors in a wall, then the wall itself will also need to be fire-resisting. (See Appendix B1 for more information about fire-resisting walls.) If a wall or floor is required to be fire-resisting then you should not make any holes in it, e.g. for extra doors or pipe ducts, without consulting a competent person.

Smoke control

In larger buildings and some units in shopping complexes, there may be some form of automatic smoke ventilation (Figure 18) provided for the safety of the occupants and to assist firefighting (e.g. Smoke and Heat Exhaust Ventilation Systems (SHEVS)). These systems are designed to restrict the spread of fire and smoke usually by venting the heat and smoke through the roof or via other routes to the outside. Low level inlet air is essential for the operation of SHEVS and all openings for this purpose should not be obstructed.

Special down-stands may have been installed to create a reservoir which will contain the smoke and hot gases at roof level, while vents allow the smoke to escape. It is important that any smoke can flow easily into the reservoirs and that nothing which could cause an obstruction, e.g. display material, is fixed near the vents. It is important that any smoke can flow easily into the reservoirs and that nothing which could cause an obstruction, e.g. large advertising displays, is fixed near the vents.

In shopping complexes the front of individual shops often forms part of the smoke control design. If your building has smoke vents fitted, or any other form of smoke control, then you may need to seek advice from someone who is competent in such systems.

Further information on smoke control can be found from CIBSE Guide E[64] or from the BRE.[65]

Figure 18: Smoke vents

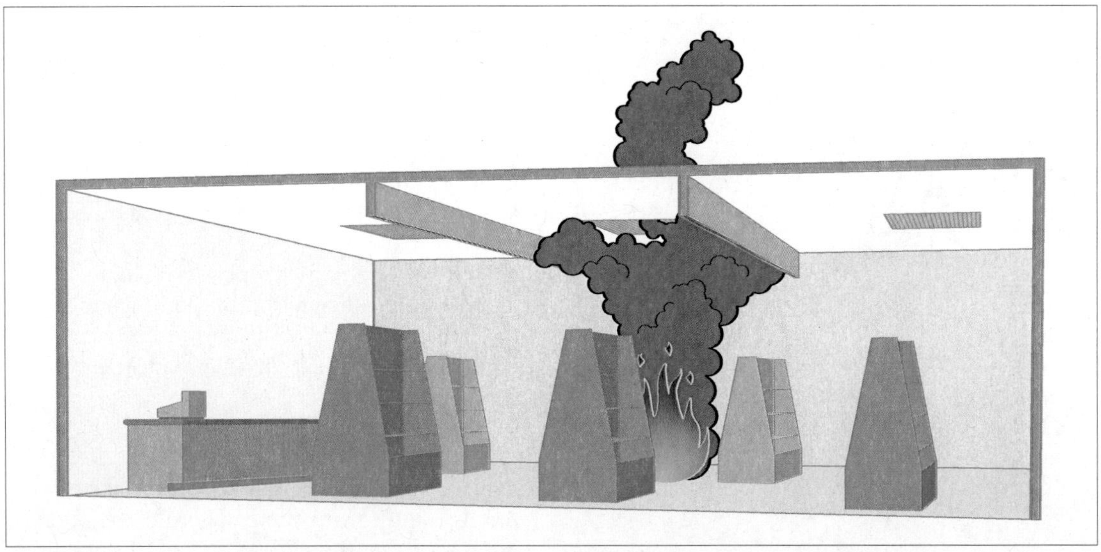

1.12 Arson

Recent studies indicate that, across all premises types, over 2,100 deliberately set fires, resulting in two deaths and 55 injuries, occur every week. In shops it is estimated that 44% of all fires are deliberately set.* All premises can be targeted either deliberately or just because they offer easy access.

Be aware of other small, deliberately set fires in the locality, which can indicate an increased risk to your premises. Be suspicious of any small 'accidental' fires on the premises and investigate them fully and record your findings.

Fires started deliberately can be particularly dangerous because they generally develop much faster and may be intentionally started in escape routes. Of all the risk-reduction measures, the most benefit may come from efforts to reduce the threat from arson.

Measures to reduce arson may include the following:

- ensure the outside of the premises is well lit and, if practical, secure the perimeter of the premises;
- thoroughly secure all entry points to the premises, including windows and the roof, but make sure that this does not compromise people's ability to use the escape routes;
- make sure you regularly remove all combustible rubbish;
- do not place rubbish skips adjacent to the building and secure waste bins in a compound separated from the building;
- do not park vehicles or store goods or materials in the open next to windows or doors opening into buildings;
- encourage staff to report people acting suspiciously;
- remove automatic entry rights from staff who have been dismissed;
- ensure that your security alarm/fire-detection system is monitored and acted on;
- secure flammable liquids so that intruders cannot use them;
- secure all storage areas and unused areas of the building that do not form part of an escape route against unauthorised access, ensure access to keys to those areas is restricted; and
- fit secure metal letterboxes on the inside of letter flaps to contain any burning materials that may be pushed through.

Further guidance on reducing the risk of arson has been published by the Arson Prevention Bureau.**

1.13 Help for people with special needs

Of all the people who may be especially at risk you will need to pay particular attention to people who may have special needs, including those with a disability. The Disability Rights Commission estimates that 11 million people in this country have some form of disability, which may mean that they find it more difficult to leave a building if there is a fire. Under the Disability Discrimination Act,[13] if disabled people could realistically expect to use your premises, then you must anticipate any reasonable adjustments that would make it easier for that right to be exercised.

The Disability Discrimination Act[13] includes the concept of 'reasonable adjustments' and this can be carried over into fire safety law. It can mean different things in different circumstances. For a small business, it may be considered reasonable to provide contrasting colours on a handrail to help those with vision impairment to follow an escape route more easily. However, it might be unreasonable to expect that same business to install an expensive voice-alarm system. Appropriate 'reasonable adjustments' for a large business or organisation may be much more significant.

If disabled people are going to be in your premises then you must also provide a safe means for them to leave if there is a fire. You and your staff should be aware that disabled people may not react, or can react differently, to a fire warning or small fire. You should give similar consideration to others with special needs such as parents with young children or the elderly.

* Fire Statistics UK, 2004, ODPM: 2006.

** Visit www.arsonpreventionbureau.org.uk for more information.

In premises with a simple layout, a common-sense approach, such as offering to help lead a blind person or helping an elderly person down steps may be enough. In more complex premises, more elaborate plans and procedures will be needed, with trained staff assigned to specified duties. In this case you may also wish to contact a professional consultant or take advice from disability organisations.

Consider the needs of those with mental disabilities or spatial recognition problems. The range of disabilities encountered can be considerable, extending from mild epilepsy to complete disorientation in an emergency situation. Many of these can be addressed by properly trained staff, discreet and empathetic use of the 'buddy system' or by careful planning of colour and texture to identify escape routes.

Where people with special needs use or work in the premises, their needs should, so far as is practicable, be discussed with them. These will often be modest and may require only changes or modifications to existing procedures. You may need to develop individual 'personal emergency evacuation plans' (PEEPs) for disabled people who frequently use a building. They will need to be confident of any plan/PEEP that is put in place after consultation with them. As part of your consultation exercise you will need to consider the matter of personal dignity.

If members of the public use your building then you may need to develop a range of standard PEEPs which can be provided on request to a disabled person or others with special needs.

Guidance on removing barriers to the everyday needs of disabled people is in BS 8300.[14] Much of this advice will also help disabled people during an evacuation.

Further advice can be obtained from the Disability Rights Commission at www.drc-gb.org.

Section 2 Further guidance on fire detection and warning systems

Where an electrical fire-warning system is necessary then a straightforward arrangement typically includes the following:

- manual call points (break-glass call points) next to exits with at least one call point on each floor;
- electronic sirens or bells; and
- a control and indicator panel.

An alternative system of interconnected combined manual call points and sounders may be acceptable.

If your building has areas where a fire could develop undetected or where people work alone and might not see a fire, then it may be necessary to upgrade your fire-warning system to incorporate automatic fire detection or install an automatic fire-detection and warning system.

If, for any reason, your system fails you must still ensure that people in your premises can be warned and escape safely. A temporary arrangement, such as gongs, whistles or air horns, combined with suitable training, may be acceptable for a short period, pending system repairs.

The fire warning sound levels should be loud enough to alert everyone, taking into account background noise. In areas with high background noise, or where people may be wearing hearing protectors, the audible warning should be supplemented, e.g. with visual alarms.

People with hearing difficulties
Where people have hearing difficulties, particularly those who are profoundly deaf, then simply hearing the fire warning is likely to be the major difficulty. If these persons are never alone while on the premises then this may not be a serious problem, as it would be reasonable for other occupants to let them know that the building should be evacuated. If a person with hearing difficulties is likely to be alone, then consider other means of raising the alarm. Among the most popular are visual beacons and vibrating devices or pagers that are linked to the existing fire alarm.

Voice alarms
Research has shown that some people and, in particular, members of the public, do not always react quickly to a conventional fire alarm. Voice alarms are therefore becoming increasingly popular and can also incorporate a public address facility. The message or messages sent must be carefully considered. It is therefore essential to ensure that voice-alarm systems are designed and installed by a person with specialist knowledge of these systems.

Schematic plan
In order to quickly determine where a fire has been detected, you should consider displaying a schematic plan showing fire alarm zones in a multi-zoned system adjacent to the control panel.

2.1 Manual call points

Manual call points, often known as 'break-glass' call points, enable a person who discovers a fire to immediately raise the alarm and warn other people in the premises of the danger.

People leaving a building because of a fire will normally leave by the way they entered. Consequently, manual call points are normally positioned at exits and storey exits that people may reasonably be expected to use in case of fire, not just those designated as fire exits. However it is not necessary in every case to provide call points at every exit.

Manual call points should normally be positioned so that, taking into account all fixtures and fittings, machinery and stock are in place, no one should have to travel more than 45m to the nearest alarm point. This distance may need to be less if your premises cater for people of limited mobility or there are particularly hazardous areas. They should be conspicuous (red), fitted at a height of about 1.4m (or less for premises with a significant number of wheelchair users), and not in an area likely to be obstructed.

2.2 Automatic fire detection

Automatic fire detection may be needed for a number of reasons. These can include:

- if you have areas where people are isolated or remote and could become trapped by a fire because they are unaware of its development, such as lone workers;
- if you have areas where a fire can develop unobserved (e.g. storerooms);
- as a compensating feature, e.g. for inadequate structural fire protection, in dead-ends or where there are extended travel distances; and
- where smoke control and ventilation systems are controlled by the automatic fire-detection system.

If you have an automatic fire detection system, the system should:

- be designed to accommodate the emergency evacuation procedure;
- give an automatic indication of the fire warning and its location. If the indicator panel is located in a part of the premises other than the control point (for example, to the secretary's office) there should ideally be a repeater panel sited in the control point;
- be maintained and tested by a competent person; and
- communicate with a central control room (if you have one).

New automatic fire detection systems should be designed and installed by a competent person. Further guidance is given in BS 5839-1[16] or a more recent standard where applicable.

Where the public address system is part of the fire warning system it should be connected to an auxiliary power source to ensure the continued use of the system in the event of fire or other emergency.

Whichever warning or detection systems are in place, however, if a fire occurs the fire and rescue service should always be called immediately.

2.3 Reducing false alarms

False alarms from automatic fire detection systems are a major problem and result in many unwanted calls to the fire and rescue service every year. Guidance on reducing false alarms has been published by ODPM/CFOA/BFPSA.[15]

If there are excessive false alarms in your premises, people may become complacent and not respond correctly to a warning of a real fire. In such circumstances, you may be failing to comply with fire safety law. All false alarms should be investigated to identify the cause of the problem and remedial action taken.

To help reduce the number of false alarms, the system design and location of detection and activation devices should be reviewed against the way the premises are currently used. For example, if a store room has been converted to staff area with cooking facilities (e.g. a microwave and toaster) then the likelihood of the detector being set off is increased. Consequently, subject to the outcome of the fire risk assessment a heat detector may be more appropriate. Similarly, if a manual call point is placed in a storage area where there is continual movement of stock, the call point is likely to be accidentally damaged. In this case a simple, fabricated hinged metal guard around the call point is likely to solve the problem.

Occasionally people set off a manual call point in the genuine, but incorrect belief that there is a fire. Nothing should be done to discourage such actions and the number of false alarms generated this way is not significant.

Further detailed guidance on reducing false alarms is available in BS 5839-1.[16]

2.4 Staged fire alarms

In the vast majority of premises sounding the fire warning system should trigger the immediate and total evacuation of the building. However, in some large or complex premises this may not be necessary as alternative arrangements may be in place.

These alternative arrangements broadly fall into two groups. Firstly, those people potentially most at risk from a fire, usually those closest to where the alarm was activated, will be immediately evacuated, while others in the building are given an alert signal and will only evacuate if it becomes necessary. This is generally called a phased evacuation and the initial movement, depending on the layout and configuration of the premises, can be either horizontal or vertical.

The second alternative is for the initial alert signal to be given to certain staff, who then carry out pre-arranged actions to help others to evacuate more easily. It requires able, fully-trained staff to be available at all times and should not be seen as a simple means of reducing disruption to working practices. Where staged alarms are being used, disabled people should be alerted on the first stage to give them the maximum time to escape.

These arrangements both require fire-warning systems capable of giving staged alarms, including an 'alert signal' and a different 'evacuate signal' and should only be considered after consultation with specialist installers and, if necessary, the relevant enforcing authority.

Such systems also require a greater degree of management input to ensure that staff and others are familiar with the system and action required.

2.5 Testing and maintenance

Your fire-warning and/or detection system should be supervised by a named responsible person, given enough authority and training to manage all aspects of the routine testing and scrutiny of the system.

The control and indicating equipment should be checked at least every 24 hours to ensure there are no specific faults. All types of fire-warning systems should be tested once a week. For electrical systems a manual call point should be activated (using a different call point for each successive test), usually by inserting a dedicated test key (Figure 19). This will check that the control equipment is capable of receiving a signal and in turn, activating the warning alarms. Manual call points may be numbered to ensure they are sequentially tested.

Testing and maintenance of the system should be carried out by a competent person.

Figure 19: Using a test key

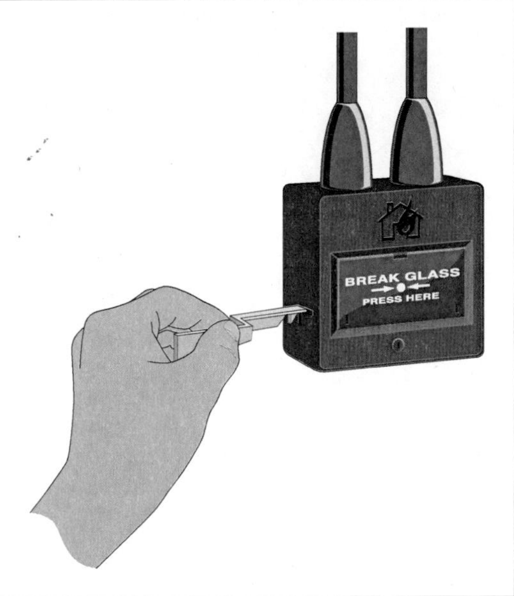

It is good practice to test the alarm at the same time each week, but additional tests may be required to ensure that staff or people present outside normal working hours are given the opportunity to hear the alarm.

Where systems are connected to a central monitoring station, arrangements should be made prior to testing to avoid unwanted false alarms. Further guidance on testing and maintenance of fire warning systems can be found in BS 5839-1.[16]

Six-monthly servicing and preventive maintenance should be carried out by a competent person with specialist knowledge of fire-warning and automatic detection systems. This task is normally fulfilled by entering into a service contract with a specialist fire alarm company.

It is good practice to record all tests, false alarms and any maintenance carried out.

2.6 Guaranteed power supply

If your fire risk assessment concludes that an electrical fire-warning system is necessary, then the Health and Safety (Safety Signs and Signals) Regulations 1996[5] requires it to have a back-up power supply.

Whatever back-up system is used, it should normally be capable of operating the fire-warning and detection system for a minimum period of 24 hours and sounding the alarm signal in all areas for 30 minutes.

2.7 New and altered systems

Guidance on the design and installation of new systems and those undergoing substantial alterations is given in BS 5839.[16] If you are unsure that your existing system is adequate you will need to consult a competent person.

Section 3 Further guidance on firefighting equipment and facilities

You have responsibility for the provision of appropriate firefighting equipment. It is also your responsibility to check that all firefighting equipment is in the correct position and in satisfactory order before the premises are used.

Appropriate staff should be trained in the use of all such equipment.

3.1 Portable firefighting equipment

Fires are classed according to what is burning. Fire extinguishers provided should be appropriate to the classes of fire found in your premises in accordance with Table 1.

Table 1: Class of fire

Class of fire	Description
Class A	Fires involving solid materials such as wood, paper or textiles.
Class B	Fires involving flammable liquids such as petrol, diesel or oils.
Class C	Fires involving gases.
Class D	Fires involving metals.
Class F	Fires involving cooking oils such as in deep-fat fryers.

Note:

1. If there is a possibility of a fire in your premises involving material in the shaded boxes then you should seek advice from a competent person.
2. It is not safe to fight fires involving aerosols with fire extinguishers.

Number and type of extinguishers

Typically for the Class A fire risk, the provision of one water-based extinguisher for approximately every 200m^2 of floor space, with a minimum of two extinguishers per floor, will normally be adequate.

Where it is determined that there are additionally other classes of fire risk, the appropriate type, number and size of extinguisher should be provided. Further information is available in BS 5306-8.[18]

Where the fire risk is not confined to a particular location, e.g. Class A fires, the fire extinguishers should be positioned on escape routes, close to the exit from the room or floor, or the final exit from the building. Similarly, where the particular fire risk is specifically located, e.g. flammable liquids, the appropriate fire extinguisher should be near to the hazard, and located so that it can be safely used. They should be placed on a dedicated stand or hung on a wall at a convenient height so that employees can easily lift them off (at about 1m for larger extinguishers, 1.5m for smaller ones, to the level of the handle). Ideally no one should have to travel more than 30m to reach a fire extinguisher. If there is a risk of malicious use you may need to use alternative, and more secure, locations.

Consider the implications of the Manual Handling Operations Regulations 1992[17] when selecting and siting firefighting equipment.

Where there are self-contained small premises, multi-purpose extinguishers which can cover a range of risks may be appropriate. Depending on the outcome of your fire risk assessment, it may be possible to reduce this to one extinguisher in very small premises with a floor space of less than 90m^2.

Extinguishers manufactured to current standards (BS EN 3-7)[79] are predominately red but may have a colour-coded area, sited above or within the instructions, denoting the type of extinguisher. Most older extinguishers, manufactured to previous standards, have bodies painted entirely in a single colour which denotes the type of extinguisher. These older extinguishers remain acceptable until they are no longer serviceable. However, it is good practice to ensure that old and new style extinguishers are not mixed on the same floor of a building.

The following paragraphs describe the different types of extinguisher. The colour referred to is the colour of the extinguisher or the colour-coded area.

Water extinguishers (red)

This type of extinguisher can only be used on Class A fires. They allow the user to direct water onto a fire from a considerable distance. A 9-litre water extinguisher can be quite heavy and some water extinguishers with additives can achieve the same rating, although they are smaller and therefore considerably lighter. This type of extinguisher is not suitable for use on live electrical equipment.

Water extinguishers with additives (red)

This type of extinguisher is suitable for Class A fires. They can also be suitable for use on Class B fires and where appropriate, this will be indicated on the extinguisher. They are generally more efficient than conventional water extinguishers.

Foam extinguishers (cream)

This type of extinguisher can be used on Class A or B fires and is particularly suited to extinguishing liquid fires such as petrol and diesel. They should not be used on free-flowing liquid fires unless the operator has been specially trained, as these have the potential to rapidly spread the fire to adjacent material. This type of extinguisher is not suitable for deep-fat fryers or chip pans.

Powder extinguishers (blue)

This type of extinguisher can be used on most classes of fire and achieve a good 'knock down' of the fire. They can be used on fires involving electrical equipment but will almost certainly render that equipment useless. Because they do not cool the fire appreciably it can re-ignite. Powder extinguishers can create a loss of visibility and may affect people who have breathing problems, and are not generally suitable for enclosed spaces.

Carbon dioxide extinguishers (black)

This type of extinguisher is particularly suitable for fires involving electrical equipment as they will extinguish a fire without causing any further damage (accept in the case of some electronic equipment e.g. computers). As with all fires involving electrical equipment, the power should be disconnected if possible.

Class 'F' extinguishers
This type of extinguisher is particularly suitable for commercial catering establishments with deep-fat fryers.

Selection, installation and maintenance of portable fire extinguishers
All portable fire extinguishers will require periodic inspection, maintenance and testing. Depending on local conditions such as the likelihood of vandalism or the environment where extinguishers are located, carry out brief checks to ensure that they remain serviceable. In normal conditions a monthly check should be enough. Maintenance by a competent person should be carried out annually.

New fire extinguishers should comply with BS EN 3-7.[79] Guidance on the selection and installation of fire extinguishers can be found in BS 5306-8,[18] for maintenance, BS 5306-3[19] and for colour coding in BS 7863.[20]

Fire blankets
Fire blankets should be located in the vicinity of the fire hazard they are to be used on, but in a position that can be safely accessed in the event of a fire. They are classified as either light duty or heavy duty. Light-duty fire blankets are suitable for dealing with small fires in containers of cooking oils or fats and fires involving clothing. Heavy-duty fire blankets are for industrial use where there is a need for the blankets to resist penetration by molten materials.

3.2 Fixed firefighting installations

These are firefighting systems which are normally installed within the structure of the building. They may already be provided in your premises or you may be considering them as a means of protecting some particularly dangerous or risk-critical area as part of your risk-reduction strategy.

Hose reels
Permanent hose reels (Figure 20) installed in accordance with the relevant British Standard (BS EN 671-3: 2000[21]) provide an effective firefighting facility. They may offer an alternative, or be in addition to, portable firefighting equipment. A concern is that untrained people will stay and fight a fire when escape is the safest option. Where hose reels are installed, and your fire risk assessment expects relevant staff to use them in the initial stages of a fire, they should receive appropriate training.

Note: It is not safe to fight fires involving aerosols with hose reels.

Maintenance of hose reels includes visual checks for leaks and obvious damage and should be carried out regularly. More formal maintenance checks should be carried out at least annually by a competent person.

Figure 20: Hose reel

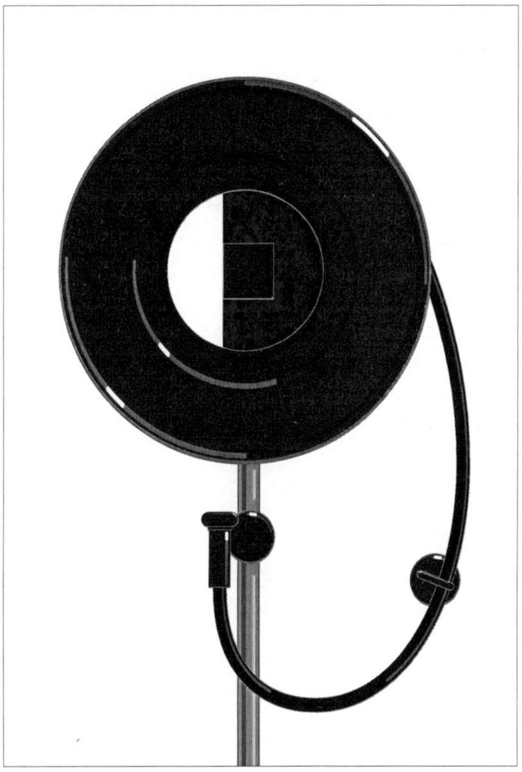

Sprinkler systems
Sprinkler systems can be very effective in controlling fires. They can be designed to protect life and/or property and may be regarded as a cost-effective solution for reducing the risks created by fire. Where installed, a sprinkler system is usually part of a package of fire precautions in a building and may form an integral part of the fire strategy for the building.

Sprinkler protection could give additional benefits, such as a reduction in the amount of portable firefighting equipment necessary, and the relaxation of restrictions in the design of buildings.

Guidance on the design and installation of new sprinkler systems and the maintenance of all systems is given in the Loss Prevention Council (LPC) Rules or BS EN 12845[22] or BS 5306-2[60] and should only be carried out by a competent person.

Routine maintenance by on-site personnel may include checking of pressure gauges, alarm systems, water supplies, any anti-freezing devices and automatic booster pump(s).

A competent maintenance contractor should provide guidance on what records need to be completed.

Following a sprinkler operation the sprinkler system should be reinstated by a competent person. A stack of spare sprinkler bulbs should be available on site for replacements, preferably in a seperate building e.g. the pumphouse.

If a sprinkler system forms an integral part of your fire strategy it is imperative that adequate management procedures are in place to cater for those periods when the sprinkler system is not functional. This should form part of your emergency plan. Although the actual procedures will vary, such measures may include the following:

- Restore the system to full working order as soon as possible.
- Limit any planned shutdown to low-risk periods when numbers of people are at a minimum (e.g. at night) or when the building is not in use. This is particularly important when sprinklers are installed to a life safety standard or form part of the fire safety engineering requirements.
- You may need to isolate the area without the benefit of working sprinklers from the rest of the premises by fire-resisting material.
- Avoid higher-risk processes such as 'hot-work'.
- Extra staff should be trained and dedicated to conducting fire patrols.
- Any phased or staged evacuation strategy may need to be suspended. Evacuation should be immediate and complete. (Exercise caution as the stairway widths may have been designed for phased evacuation only.)
- Inform the local fire and rescue service.

If, having considered all possible measures, the risk is still unacceptable then it will be necessary to close all or part of the building. If in doubt you should seek guidance from a competent person.

Other fixed installations

There are a number of other fixed installations including water mist, gaseous, deluge and fixed powder systems. If your premises have a fixed firefighting system that you are unfamiliar with, then seek advice. Where a fixed firefighting system forms an integral part of your fire safety strategy, it should be maintained in accordance with the relevant British Standard by a competent person.

3.3 Other facilities (including those for firefighters)

Building Regulations and other Acts, including local Acts, may have required firefighting equipment and other facilities to be provided for the safety of people in the building and to help firefighters. Fire safety law places a duty on you to maintain such facilities in good working order and at all times.

These may include:

- access for fire engines and firefighters;
- firefighting shafts and lifts;
- fire suppression systems e.g. sprinklers, water mist and gaseous;
- smoke-control systems;
- dry or wet rising mains and firefighting inlets;
- information and communication arrangements e.g. fire telephones and wireless systems and information to brief the fire and rescue service when they arrive; and
- firefighters' switches.

The Workplace (Health, Safety and Welfare) Regulations 1992[23] also require that systems provided for safety within a workplace are maintained.

Access for fire engines and firefighters

Buildings that have been constructed to modern building regulations or in accordance with certain local Acts will have been provided with facilities that allow fire engines to approach and park within a reasonable distance so that firefighters can use their equipment without too much difficulty.

These facilities may consist of access roads to the building, hard standing areas for fire engines and access into the building for firefighters. It is essential that where such facilities are provided they are properly maintained and available for use at all relevant times.

Where a building is used by a number of different occupants you will need to ensure co-operation between the various 'responsible 'persons' to maintain fire and rescue service access. In exceptional cases, where access is persistently obstructed, you may need to make additional arrangements.

See Approved Document B to the Building Regulations for more information.[24]

Firefighting shafts and lifts

Firefighting shafts (Figure 21) are provided in larger buildings to help firefighters reach floors further away from the building's access point. They enable firefighting operations to start quickly and in comparative safety by providing a safe route from the point of entry to the floor where the fire has occurred.

Figure 21: Firefighting shaft

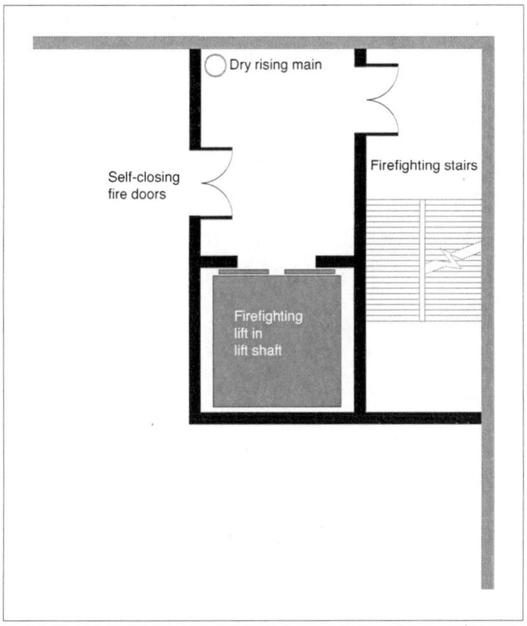

Entry points from a stairway in a firefighting shaft to a floor will be via a lobby, through two sets of fire and smoke-resisting doors and walls. Many people will use the stairway for normal movement through the building and it is important that the safety features are not compromised by doors being wedged open.

Most firefighting shafts will also incorporate a firefighting lift which opens into the lobby. The lift will have a back-up electrical supply and car control overrides. The primary function of the lift is to transport firefighting personnel and their equipment to the scene of a fire with the minimum amount of time and effort. It may also be used to help evacuate less mobile people.

Alterations that might affect the shaft should not be made without first liaising with other responsible persons, any owners or managing agents and the enforcing authority. Any proposed changes will require Building Regulation approval from a Building Control Body.

Where a firefighting shaft is provided, it should be maintained by a competent person.

Suppression systems

Fire suppression systems can include sprinklers and other types of fixed installations designed to automatically operate and suppress a fire. Such systems should be maintained by a competent person.

Smoke control systems

These are complex systems that are provided for life safety of occupants, assistance to firefighters and property protection by clearing hot smoke and gases from the building. If you have one of these systems provided in your premises you should ensure you understand how it operates and that it is maintained in full working order. If your system is part of a larger system then you should liaise with other occupiers and building managers.

The smoke control system should be maintained by a competent person who is familiar with the fire engineering performance specifications of that specific system.

Where these systems are installed in addition to a sprinkler system, then the design and installation of each system should not act detrimentally on one another. A competent person should be employed to confirm this.

Dry and wet rising fire mains

The rising fire main (Figure 22) is an important facility for the fire and rescue service in taller buildings. It consists of an inlet box where firefighters can connect their hoses, a pipe running up or through the building, outlet valves on each floor level and an air vent at the top.

Figure 22: Rising main

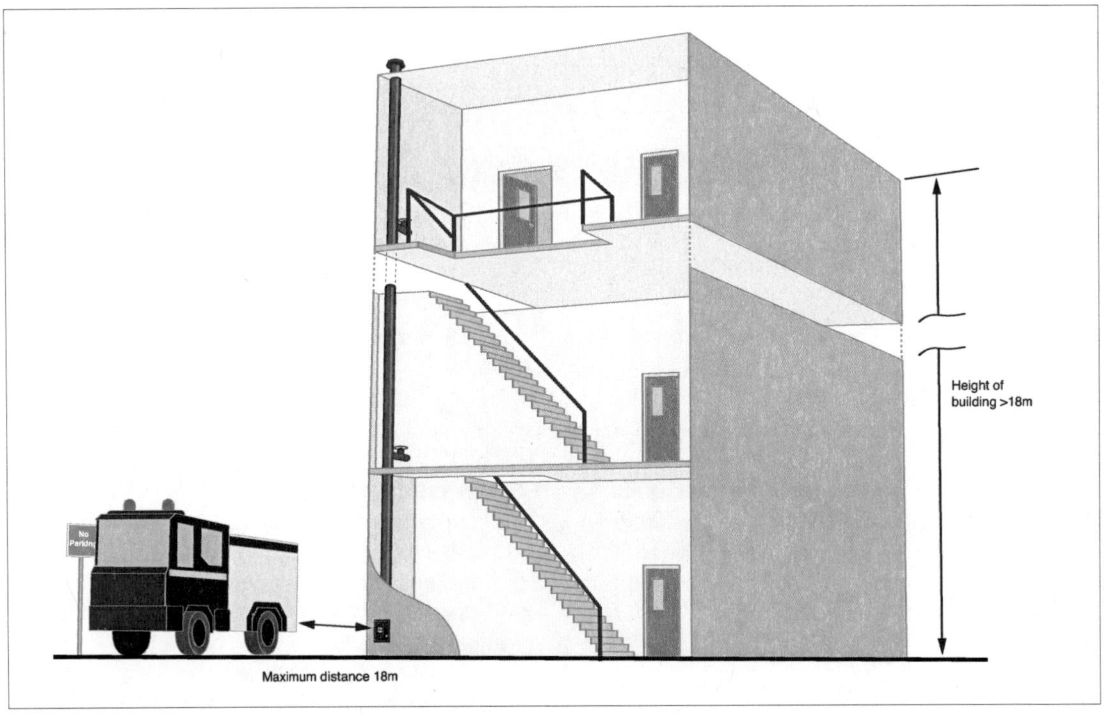

It is important that fire mains remain in good working order. Issues to be considered can include the following:

- The physical approach to the inlet box should be such that a fire engine can park within 18 m with the inlet box in view.
- Prohibit car parking in front of the inlet box.
- Secure the inlet box in such a way that firefighters can open the door without too much difficulty.
- It is advisable to lock the landing valves in the closed position, usually with a leather strap and padlock.

Foam inlets

Foam intlets are special inlets are usually fitted to provide an efficient way of extinguishing a fire in a basement or other area of high risk such as plant room. In many respects they look the same as rising main inlet boxes, but the door should be clearly marked 'foam inlet'. The risk area should be kept clear of obstructions to allow the foam to spread into the compartment.

Maintenance of rising mains and foam inlets

All types of rising mains and foam inlets together with associated valves should be maintained and tested on a regular basis by a competent person. Although there are no recommended periods between maintenance checks it would be prudent to carry out an annual service.

Firefighters' switches

Safety switches are normally provided to isolate high voltage luminous signs or to cut off electrical power. In the case of existing installations, if they have been provided in accordance with previous legislation (e.g. the Local Government (Miscellaneous Provisions) Act 1982[25]), then it is likely that they will comply with the Order. If this is not the case, then you may need to consult the enforcing authority regarding the suitability of its location and marking. Testing should be carried out in accordance with the manufacturer's instructions. If you have no such instructions then an initial test should be carried out by a competent electrician.

Other firefighting facilities

As well as those already mentioned, other facilities to assist firefighters may have been installed in your premises, and should be properly maintained by a competent person. Your maintenance audit (see Appendix A for an example checklist) should include these. Such facilities can include:

- information signs for firefighters;
- static water supplies, private hydrants, meter bypass valves and underground tanks;
- standby fire pumps, electrical generators, air pumps and hydraulic motors; and
- manual/self-closing devices for roller shutter doors in fire compartment walls.

Section 4 Further guidance on escape routes

Introduction

This section provides further guidance on the general principles that apply to escape routes and provides examples of typical escape route solutions for a range of common building layouts. The guidance is based on premises of normal risk so if your premises (or part of your premises) are higher (or lower) risk you should adapt the solution accordingly.

You are not obliged to adopt any particular solution for escape routes in this section if you prefer to meet the relevant requirement in some other way. If you decide to adopt some alternative arrangement it will need to achieve at least an equivalent level of fire safety.

Refer to the glossary (Appendix D) for the definitions of any terms you may not be familiar with.

Levels of risk

In order to apply the guidance in this section, you need to understand that in any fire situation, the time that people have to escape before they could become affected by the fire is limited. Providing them with sufficient time usually means that as well as having an appropriate way of detecting and giving warning in case of fire, the distance that people have to travel to make their escape to a place of reasonable or total safety must be restricted.

The travel distances which are usually appropriate for this purpose (and are suggested later in this section) vary according to the level of risk in the premises (or part of them). To check your escape routes you will need to form a judgement about the level of risk that people may be at after you have taken other risk reduction (preventative and protective) measures.

In premises where there is a likelihood of a fire starting and spreading quickly (or a fire could start and grow without being quickly detected and a warning given) and affect the escape routes before people are able to use them, then the risk should normally be regarded at 'higher'. Such premises could include those where significant quantities of flammable materials are used or stored; ready sources of ignition are present, e.g. heat producing machinery and processes; premises where significant numbers of the people present are likely to move slowly or be unable to move without assistance; and premises where the construction provides hidden voids or flues through which a fire could quickly spread.

In premises where there is a low occupancy level and all the occupants are able bodied and capable of using the means of escape without assistance; very little chance of a fire; few if any highly combustible or flammable materials or other fuels for a fire; fire cannot spread quickly; and will be quickly detected so people will quickly know that a fire has occurred and can make their escape, then the risk can usually be regarded as 'lower'.

In most cases however, the risk will usually be 'normal'.

The travel distances suggested are not hard and fast rules and should be applied with a degree of flexibility according to the circumstances. For example, in premises where the risk might otherwise be considered 'normal' but where there are a significant number of people who move slowly or may need assistance to evacuate, it would usually be appropriate to consider this a 'higher' risk. However, where other measures are in place to mitigate this, such as the availability of extra assistance and this has been planned for in your emergency plan, it may be that the risk level can be regarded as 'normal to higher'.

Equally, in premises where the risk category would otherwise be 'lower' but for the fact that a small number of occupants may move slowly or need assistance, it may be appropriate to categorise the risk as 'normal' in these circumstances.

If you are not sure about the level of risk that remains in your premises, you should seek advice from a competent person.

4.1 General principles

Suitability of escape routes

You should ensure that your escape routes are:

- suitable;
- easily, safely and immediately usable at all relevant times;
- adequate for the number of people likely to use them;
- free from any obstructions, slip or trip hazards; and
- available for access by the emergency services.

In multi-occupied premises, escape routes should normally be independent of other occupiers, i.e. people should not have to go through another occupier's premises as the route may be secured or obstructed. Where this is not possible, then robust legal agreements should be in place to ensure their availability at all times.

All doors on escape routes should open in the direction of escape and ideally be fitted with a safety vision panel. This is particularly important if more than 60 people use them or they provide an exit from an area of high fire risk.

At least two exits should be provided if a room/area is to be occupied by more than 60 persons. This number of 60 can be varied in proportion to the risk; for a lower risk there can be a slight increase, for a higher risk, lower numbers of persons should be allowed.

Movement of persons up or down a group of not less than three steps will be so obvious to those following that they will be prepared for the change in level, but movement up or down one step is not so readily observed and may easily lead to a fall. Wherever practicable, differences of level in corridors, passages and lobbies should be overcome by the provision of inclines or ramps of gradients not exceeding 1 in 12 or steps not having less than three risers in any flight. Corridors and passages should be level for a distance of 1.5 metres in each direction from any steps.

Any mirrors situated in escape routes should be sited so that persons escaping from a fire will not be thrown into confusion by any reflected image of the route they are using, or be misled as to the direction they should take to reach fire exits

While not normally acceptable, the use of ladders, floor hatches, wall hatches or window exits may be suitable for small numbers of able-bodied, trained staff in exceptional circumstances.

Fire-resisting construction

The type and age of construction are crucial factors to consider when assessing the adequacy of the existing escape routes. To ensure the safety of people it may be necessary to protect escape routes from the effects of a fire. In older premises (see Appendix C for more information on historical properties) it is possible that the type of construction and materials used may not perform to current fire standards. Also changes of occupier and refurbishment may have led to:

- cavities and voids being created, allowing the potential for a fire to spread unseen;
- doors and hardware worn by age and movement being less likely to limit the spread of smoke;
- damaged or lack of cavity barriers in modular construction; and
- breaches in fire compartment walls, floors and ceilings created by the installation of new services, e.g. computer cabling.

Where an escape route needs to be separated from the rest of the premises by fire-resisting construction, e.g. a dead-end corridor or protected stairway (see Figures 31 and 35 on pages 74 and 78 respectively), then you should ensure the following:

- Doors (including access hatches to cupboards, ducts and vertical shafts linking floors) walls, floors and ceilings protecting escape routes should be capable of resisting the passage of smoke and fire for long enough so that people can escape from the building.
- Where suspended or false ceilings are provided, the fire resistance should extend up to the floor slab level above. For means of escape purposes a 30 minutes fire-resisting rating is normally enough.
- Cavity barriers, fire stopping and dampers in ducts are appropriately installed.

If there is any doubt about the nature of the construction of your premises, ask for advice from a competent person.

Number of people using the premises

As your escape routes need to be adequate for the people likely to use them you will need to consider how many people, including employees and the public, may be present at any one time. Where premises have been subject to building regulations approval for use as either an office or a shop, the number and width of escape routes and exits will normally be enough for the anticipated number of people using the building. In such buildings where the risk has changed or buildings were constructed before national Building Regulations it is still necessary to confirm the provision.

For offices, the maximum numbers of staff, visitors and contractors liable to be in the building at the same time will be known by the responsible person. For shops, the responsible person will normally be aware of the maximum number of people liable to be present from a personal knowledge of trading patterns. There will also be an appreciation of the use of the building by those with special needs, such as the disabled.

If you propose to make changes to the use or layout of the building which may increase the number of people, you should check the design capacity by referring to guidance given in the Building Regulations Approved Document B.[24]

> Case study
>
> The function of a shop changed from selling large items of furniture to selling music CDs. While it was a furniture shop, most of the sales area was used to display furniture, so fewer customers could be accommodated in the store and only a small number of fire exits were necessary. When the shop started to sell CDs, more floor space was available which accommodated more customers. This necessitated checking the number and width of exits resulting in the need for an additional exit.

Mobility impairment

Effective management arrangements need to be put in place for those that need help to escape.

Consider the following points:

- A refuge is a place of reasonable safety in which disabled people can wait either for an evacuation lift or for assistance up or down stairs (see Figure 23). Disabled people should not be left alone in a refuge area whilst waiting for assistance with evacuation from the building. Depending on the design and fire resistance of other elements, a refuge could be a lobby, corridor, part of a public area or stairway, or an open space such as a flat roof, balcony or similar place which is sufficiently protected (or remote) from any fire risk and provided with its own means of escape and a means of communication.
- Where refuges are provided, they should be enclosed in a fire-resisting structure which creates a protected escape route which leads directly to a place of total safety and should only be used in conjunction with effective management rescue arrangements. Your fire safety strategy should not rely on the fire and rescue service rescuing people waiting in these refuges.
- If firefighting lifts (provided in high buildings as firefighting access) are to be used for evacuation, this should be co-ordinated with the fire and rescue service as part of the pre-planned evacuation procedures.
- Normal lifts may be considered suitable for fire evacuation purposes, subject to an adequate fire risk assessment and development of a suitable fire safety strategy by a competent person.
- Since evacuation lifts can fail, having reached a refuge a disabled person should also be able to gain access to a stairway (should conditions in the refuge become untenable). An evacuation lift with its associated refuge should therefore be located adjacent to a protected stairway.
- Enough escape routes should always be available for use by disabled people. This does not mean that every exit will need to be adapted. Staff should be aware of routes suitable for disabled people so that they can direct and help people accordingly.

- Stairways used for the emergency evacuation of disabled people should comply with the requirements for internal stairs in the building regulations. Specialist evacuation chairs or other equipment may be necessary to negotiate stairs.
- Plans should allow for the careful carrying of disabled people down stairs without their wheelchairs, should the wheelchair be too large or heavy. You will need to take into account health and safety manual handling procedures in addition to the dignity and confidence of the disabled person.
- Stairlifts should not be used for emergency evacuation. Where installed in a stairway used for emergency evacuation, no parts of the lift, such as its carriage rail, should be allowed to reduce the effective width of the stairway or any other part of an emergency evacuation route.
- Where ramps are necessary for the emergency evacuation of people in wheelchairs they should be as gentle as possible. Ramps should be constructed in accordance with Approved Document M.[62]

Further guidance is available in BS 5588-8[63] and BS 5588-12.[52]

Figure 23: An example of a refuge

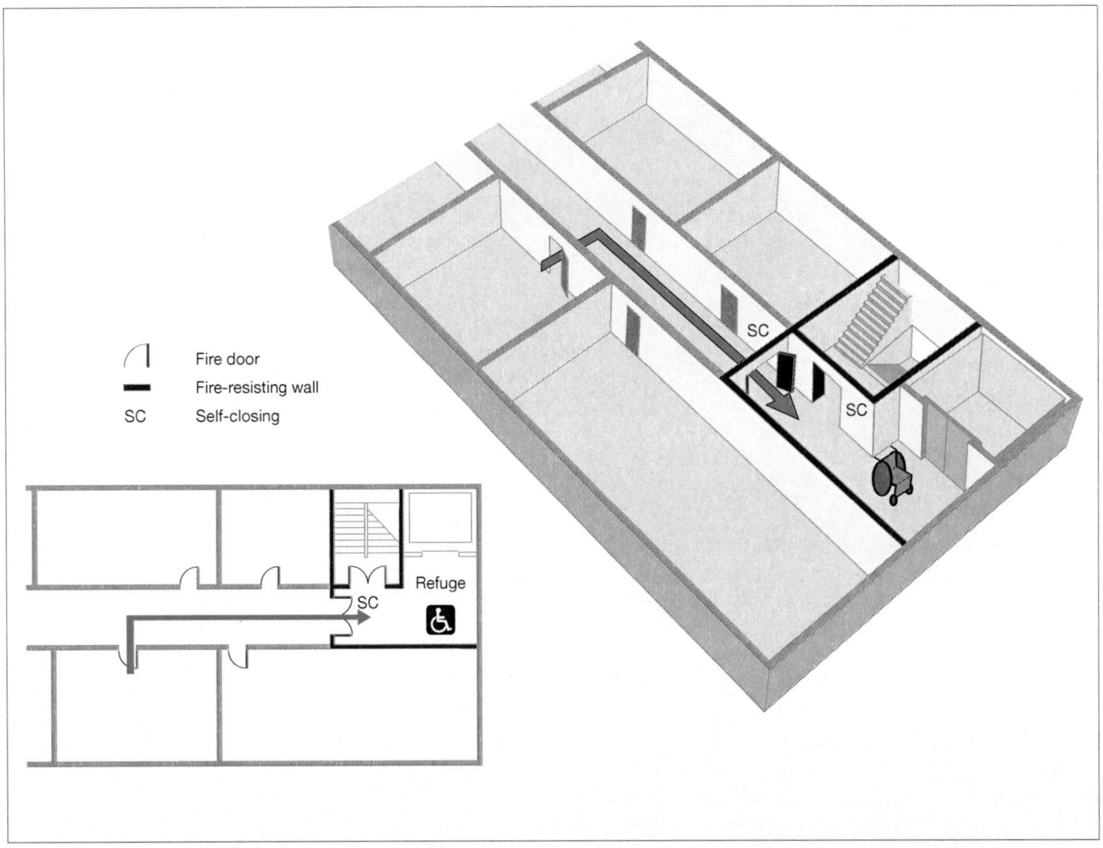

Widths and capacity of escape routes and stairways

Once you have established the maximum number of people likely to be in any part of the premises, the next step is to establish that the capacity of the escape routes is adequate for people to escape safely in sufficient time to ensure their safety in case of fire.

The capacity of a route is determined by a number of factors including the width of the route, the time available for escape and the ability of the persons using them.

The effective usable width of an escape route is the narrowest point, normally a door or other restriction such as narrowing of a corridor due to fixtures and fittings. The capacity of an escape route is measured by the number of persons per minute that can pass through it, so to establish the capacity of the route, it is first necessary to measure the width of the route at the narrowest point. The effective width of a doorway is the clear unobstructed width through the doorway when the door is open at right angles to the frame. The effective width at any other point is the narrowest clear unobstructed width through which people can pass.

The time available for escape depends on several factors. Studies of human behaviour in an emergency situation have shown that about two thirds of the time available to escape is taken up by the initial reaction to the developing situation. For example, people will decide whether the situation is real or false, often waiting to see the reaction of people around them, and generally gathering information to decide whether to act or not. The final third is taken up by the actual movement away from the area of the fire. Throughout this time the fire may be growing and spreading. Therefore to account for the limited available time for people to travel to a place of reasonable safety, the length of escape routes needs to be restricted. The suggested travel distances in this section (Table 2 on page 68) take this restriction into account.

The following guide can be used to determine the general capacities of escape routes:

A width of at least 750mm can accommodate up to:

- 80 people in higher risk premises;
- 100 people in normal risk premises; or
- 120 people in lower risk premises.

A width of at least 1050mm can accommodate up to:

- 160 people in higher risk premises;
- 200 people in normal risk premises; or
- 240 people in lower risk premises.

An additional 75mm should be allowed for each additional 15 persons (or part of 15).

Note: The minimum width of an escape route should not be less than 750mm (unless it is for use by less than five people in part of your premises) and, where wheelchair users are likely to use it, not less than 900mm.

The aggregate width of all the escape routes should be not less than that required to accommodate the maximum numbers of people likely to use them.

When calculating the overall available escape route capacity for premises that have more than one way out, you should normally assume that the widest is not available because it has been compromised by fire. If doors or other exits leading to escape routes are too close to one another you should consider whether the fire could affect both at the same time. If that is the case, it may be necessary to discount them both from your calculation.

As a general rule stairways should be at least 1050mm wide and in any case not less than the width of the escape routes that lead to them. In all cases the aggregate capacity of the stairways should be sufficient for the number of people likely to have to use them in case of fire.

Stairways wider than 2100mm should normally be devided into sections, each separated from the adjacent section by a handrail, so that each section measured between the handrails is not less than 1050mm wide.

Travel distance

Having established the number and location of people and the exit capacity required to evacuate them safely, you now need to confirm that the number and location of existing exits is adequate. This normally determined by the distance people have to travel to reach them.

Table 2 gives guidance on travel distances. It should be understood, however, that these distances are flexible and may be increased or decreased depending upon the level of risk after you have put in place the appropriate fire-prevention measures (Part 1, Step 3.3).

In new buildings which have been designed and constructed in accordance with modern building standards the travel distances will already have been calculated. Once you have completed your fire risk assessment you need to confirm that those distances are still relevant.

When assessing travel distances you need to consider the distance to be travelled by people when escaping, allowing for walking around furniture or display material etc. (see Figure 24). The distance should be measured from all parts of the premises (e.g. from the most remote part of an office or shop on any floor) to the nearest place of reasonable safety which is:

- a protected stairway enclosure (a storey exit);
- a separate fire compartment from which there is a final exit to a place of total safety; or
- the nearest available final exit.

Table 2: Suggested travel distances

Escape routes	Suggested range of travel distance
Where more than one escape route is provided	25m in higher fire-risk area[1] 45m in normal fire-risk area 60m in lower fire-risk area[2]
Where only a single escape route is provided	12m in higher fire-risk area[1] 18m in normal fire-risk area 25m in lower fire-risk area[2]

Note 1:
Where there are small high-risk areas this travel distance should apply. Where the risk assessment indicates that the whole building is high-risk, ask advice from a competent person.

Note 2:
The travel distance for lower risk premises should only be applied in exceptional cases in the very lowest risk premises where densities are low, occupants are familiar with the premises, excellent visual awareness, and very limited combustibles.

The travel distances given in Table 2 are based on those recommended in Approved Document B of the Building Regulations[24] and are intended to complement the other fire safety recommendations in Approved Document B. Your current escape route travel distances may be different from these since they may be based on recommendations made in alternative guidance.

Where your route leads to more than one final exit, but only allows initial travel in a single direction (e.g. from a room or dead end, see Figures 28 and 29), then this initial travel distance should be limited to that for a 'single escape route' in Table 2. However, your total travel distance should not exceed that for 'more than one escape route'.

Measuring travel distance

The figures that follow are schematic and are intended to represent part of a larger building.

The route taken through a room or space will be determinted by the layout of the contents e.g. work stations, aisle layout (Figure 24). It is good practice to ensure routes to the exits are kept as direct and short as possible. In a small room there may be only one exit but in a larger room or area there may be many exits.

In some cases where the contents are moved around or the space is liable to frequent change, e.g. in a storage area or where racking is moveable you should ensure that the exits, or the routes to them, do not become blocked or the length of the route is not significantly extended.

Figure 24: Measuring travel distance

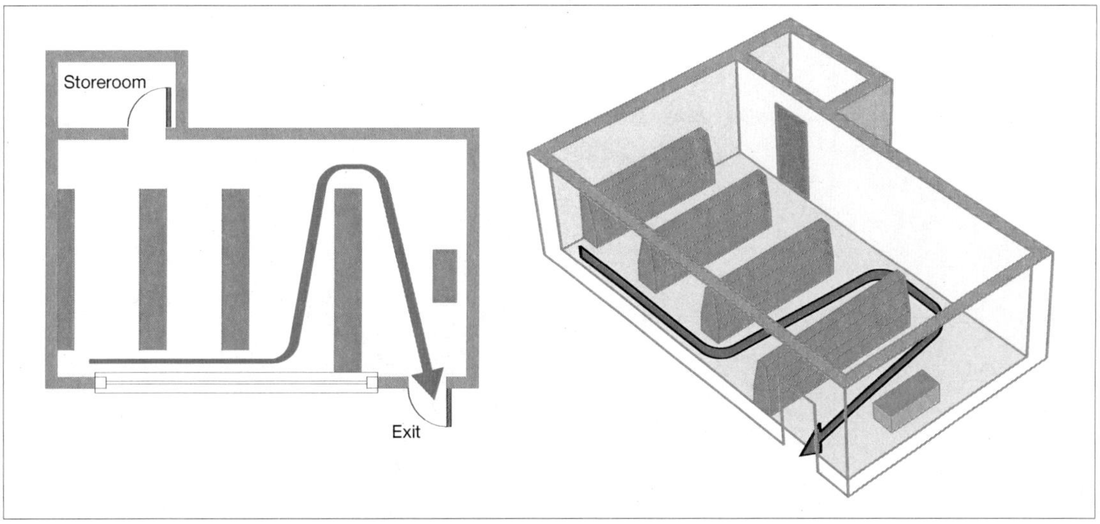

Inner rooms

Where the only way out of a room is through another room (Figure 25), an unnoticed fire in the outer room could trap people in the inner room. This means of exit should be avoided where possible. If, however, this cannot be achieved then adequate warning of a fire should be provided by **any one** of the following means:

- a vision panel between the two rooms providing adequate vision to give an indication of the conditions in the outer room and the means of escape;
- a large enough gap between the dividing wall and the ceiling, e.g. 500mm, so that smoke will be seen; or
- an automatic smoke detector in the outer room that will sound a warning in the inner room.

In addition, the following points should also be considered:

- Restrict the number of people using an inner room to 60.
- Access rooms should be under the control of the same person as the inner room.
- The travel distance from any point in the inner room to the exit from the access room should be restricted to escape in one direction only (see Table 2 on page 68), unless there are alternative exits from the access room.
- No one should have to pass through more than one access room while making their escape.
- The outer room should not be an area of high fire risk.

Figure 25: Inner rooms

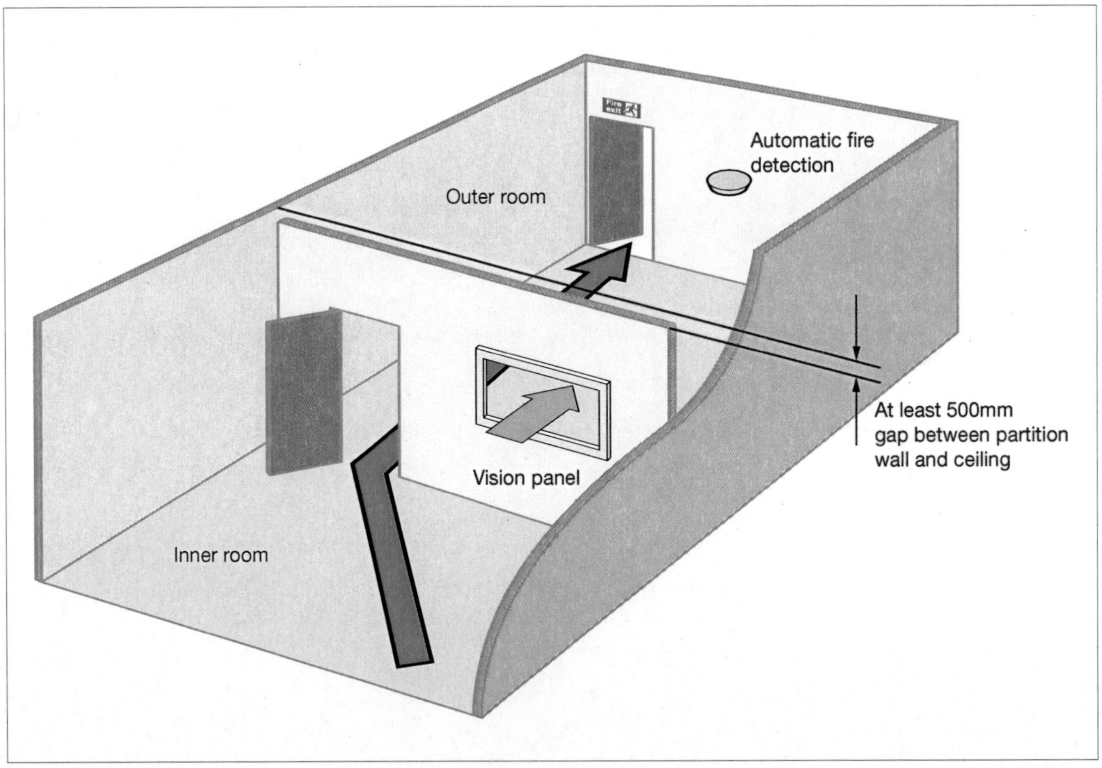

Alternative exits

Where alternative exits from a space or room are necessary, they should wherever possible be located at least 45° apart (see Figure 26) unless the routes to them are seperated by fire-resisting construction (see Figure 27). If in doubt consult a competent person.

Figure 26: Alternative exits

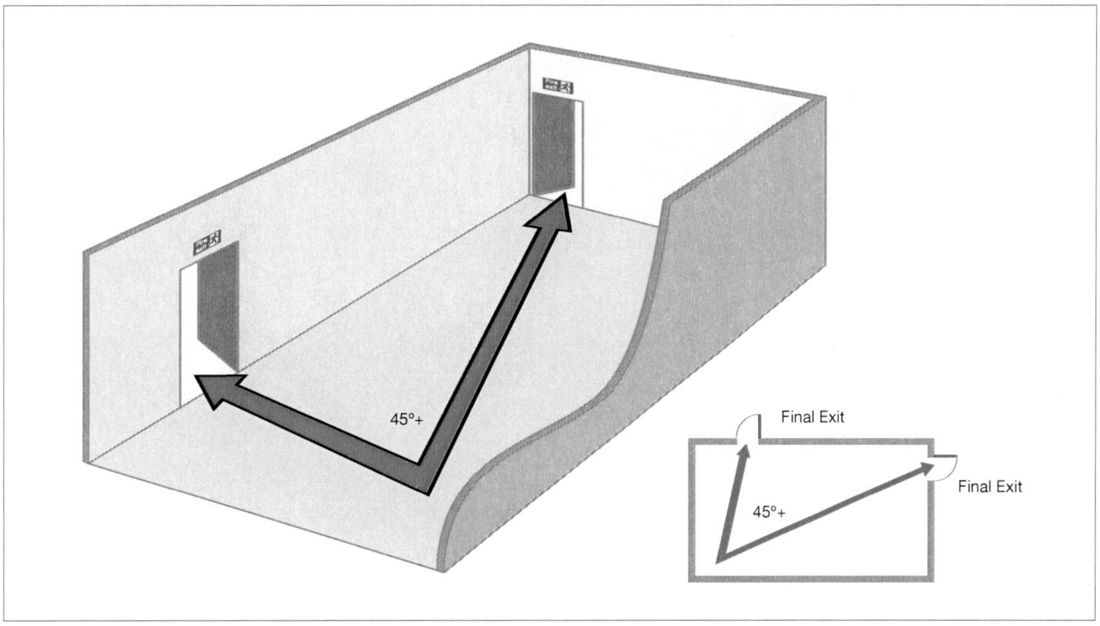

Figure 27: Alternative exits seperated by fire-resisting construction

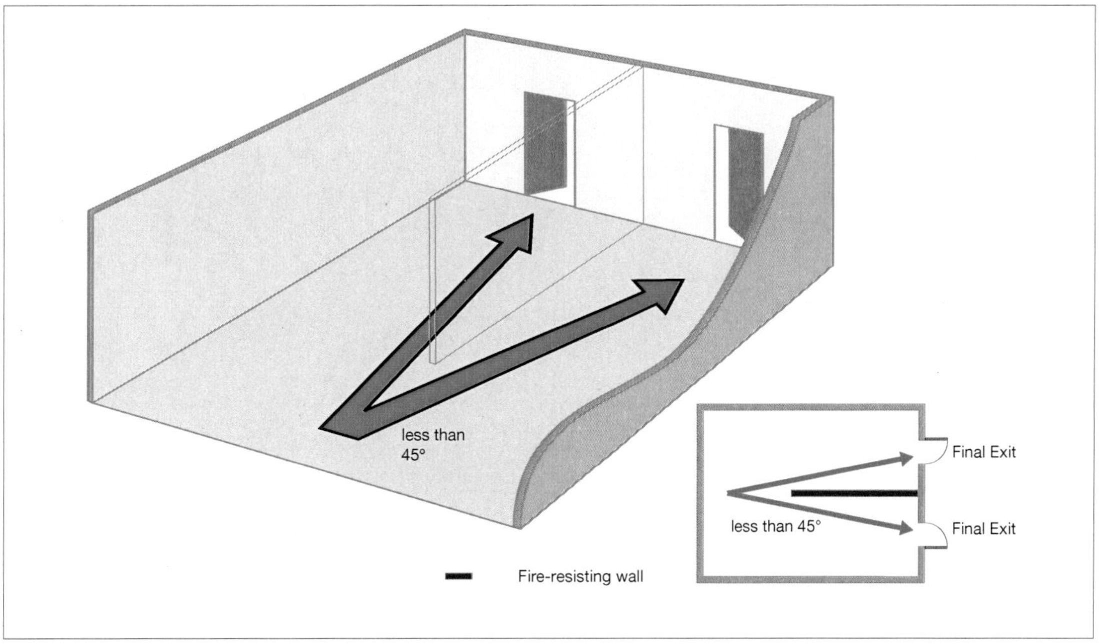

Measuring travel distances for initial dead-end travel

Where the initial direction of travel in an open area (see Figure 28) or within an inner room (see Figure 29) is in one direction only the travel distance (A–B) should be limited to that for a 'single escape route' in Table 2 on page 68.

Any alternative exits should be positioned to ensure a fire will not compromise both exits. The maximum total travel distance recommended in Table 2 should apply to the nearest exit (Figure 28 and 29, distance A–C). However, since you have two exits, your total travel distance should not exceed that for 'more than one escape route' in Table 2.

Figure 28: Measuring travel distance from initial dead end (open plan)

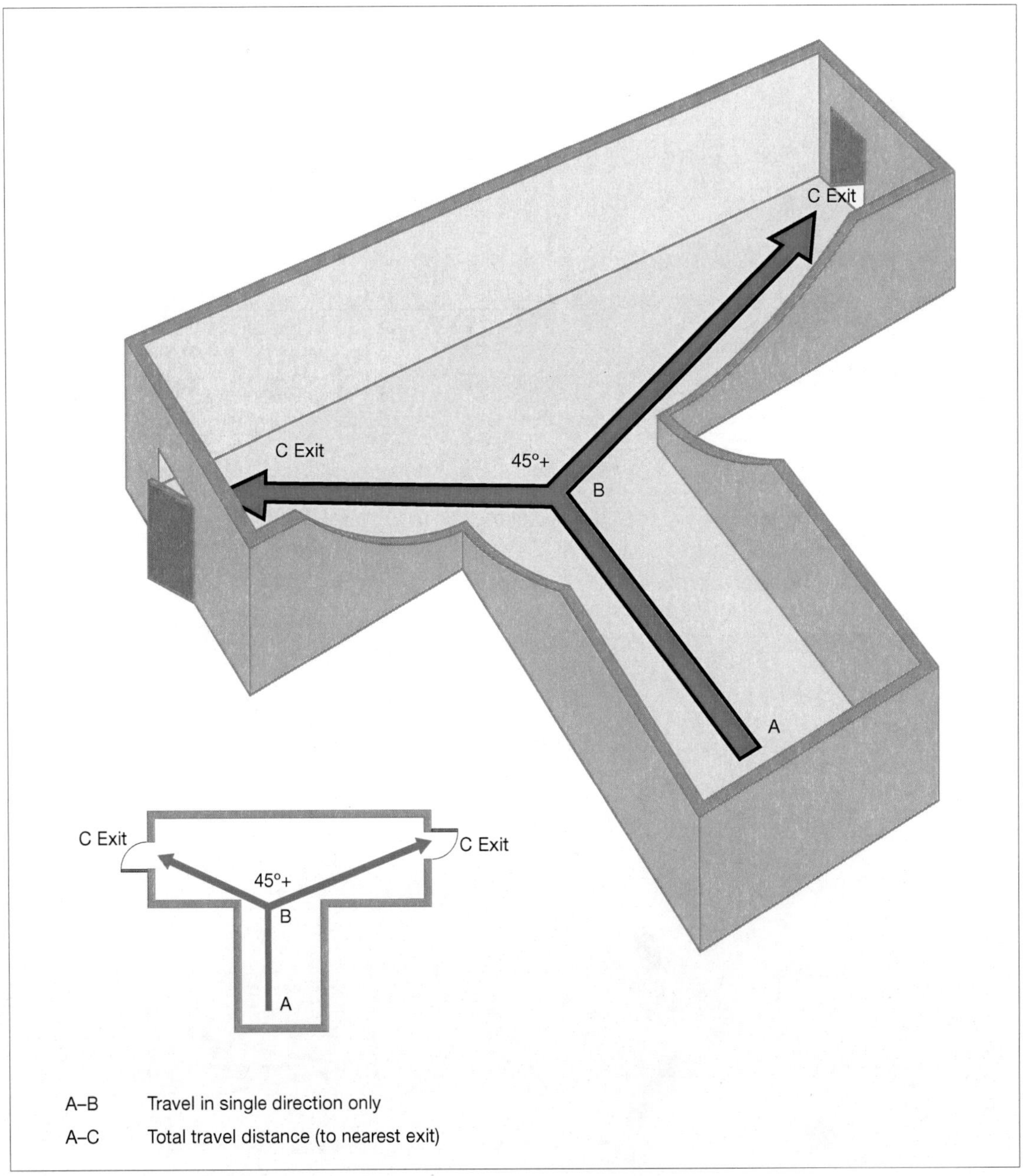

Figure 29: Measuring travel distance from initial dead end (inner room)

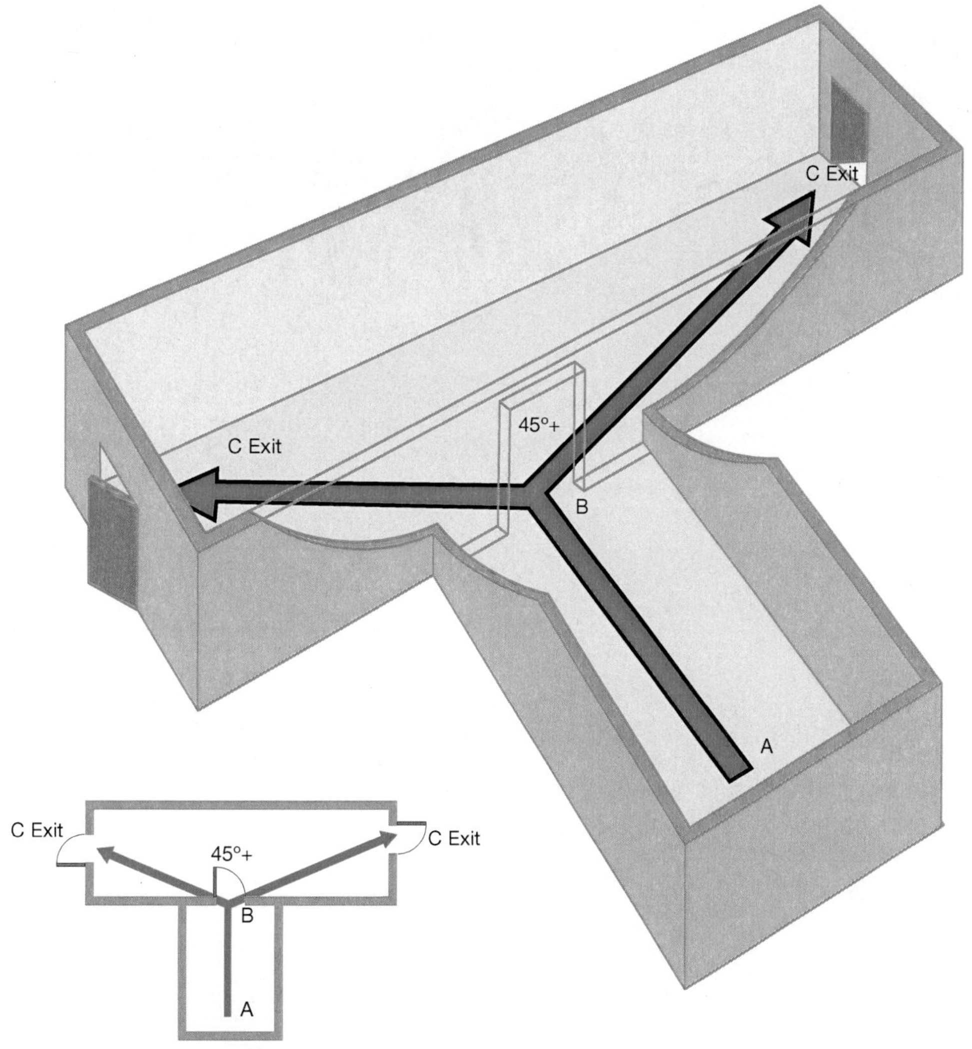

A–B Travel in single direction only

A–C Total travel distance (to nearest exit)

Note: Additional measures are necessary to give warning of fire for inner rooms (Figure 25)

Escape routes with dead-end conditions

If your premises has escape routes from which escape can be made in one direction only (a dead end), then an undetected fire in that area could affect people trying to escape. To overcome this problem, limit the travel distance (see Table 2) and use one of the following solutions:

- Fit an automatic fire detection and warning system in those areas where a fire could present a risk to the escape route (see Figure 30).
- Protect the escape route with fire-resisting construction to allow people to escape safely past a room in which there is a fire (see Figure 31).
- Provide an alternative exit (see Figure 32).

Alternative approaches may be acceptable, although expert advice may be necessary.

Figure 30: Dead-end condition with automatic fire detection

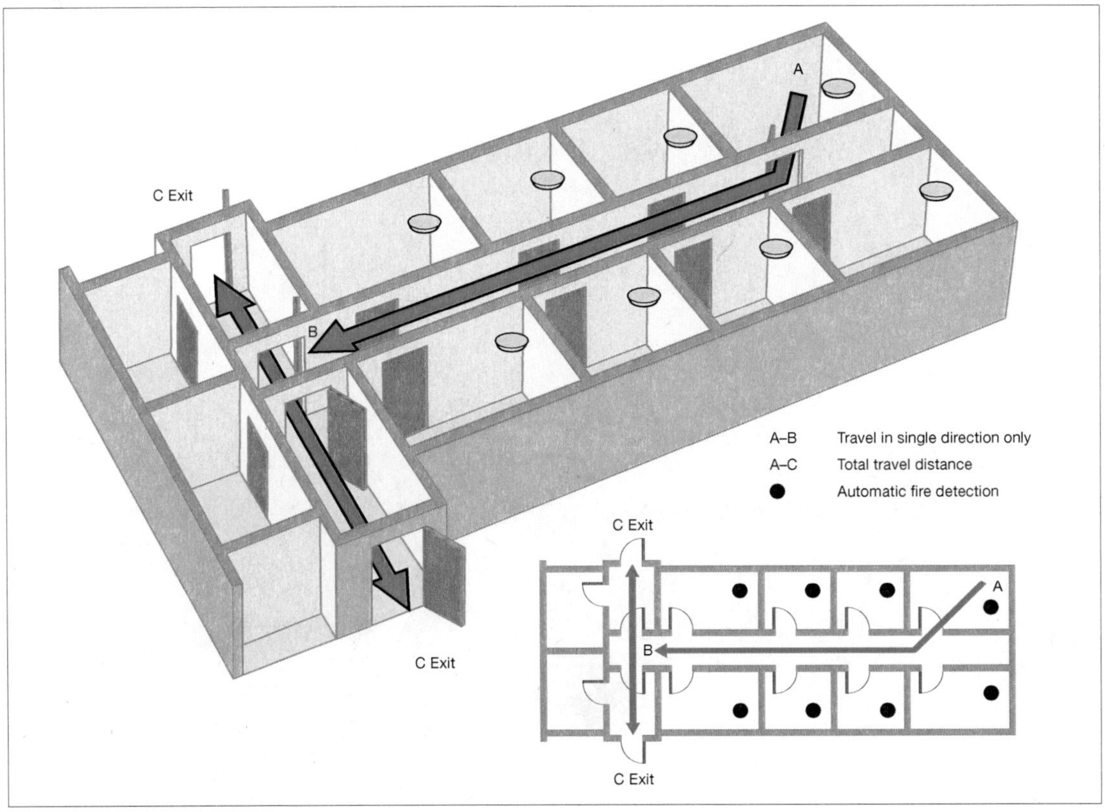

Figure 31: Dead-end condition with fire-resisting construction

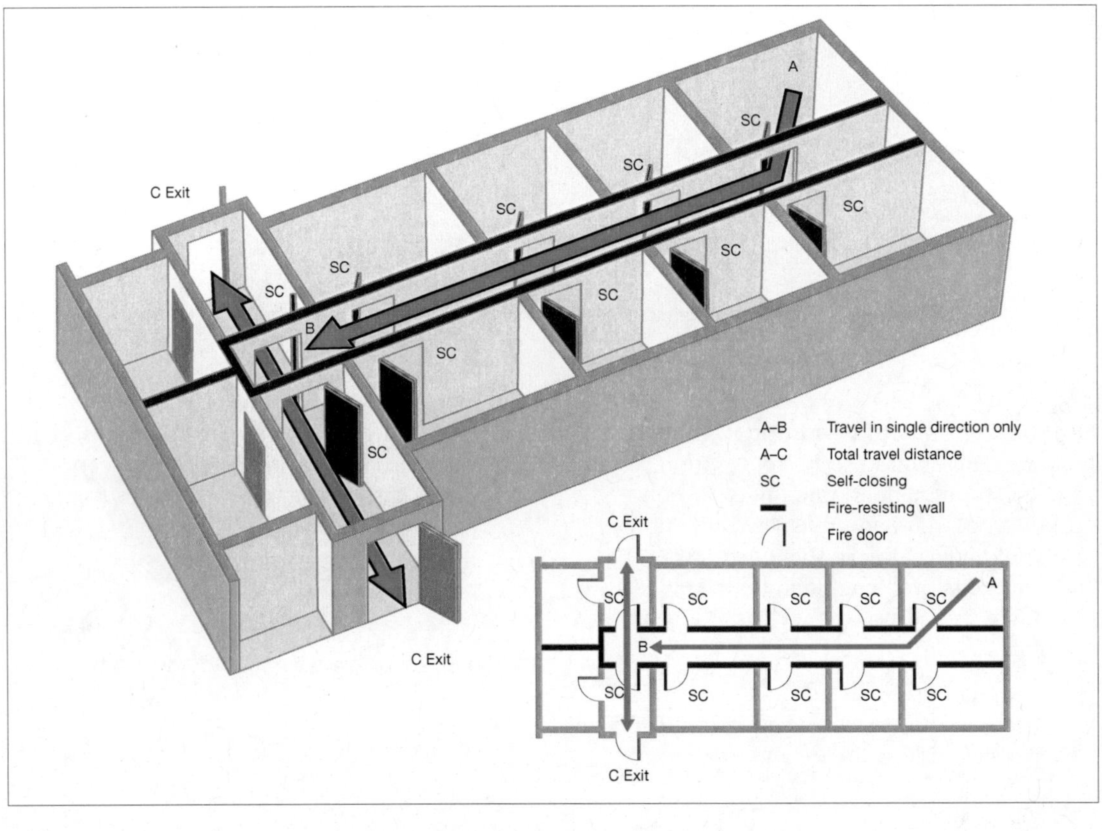

Figure 32: Dead-end condition provided with an alternative exit

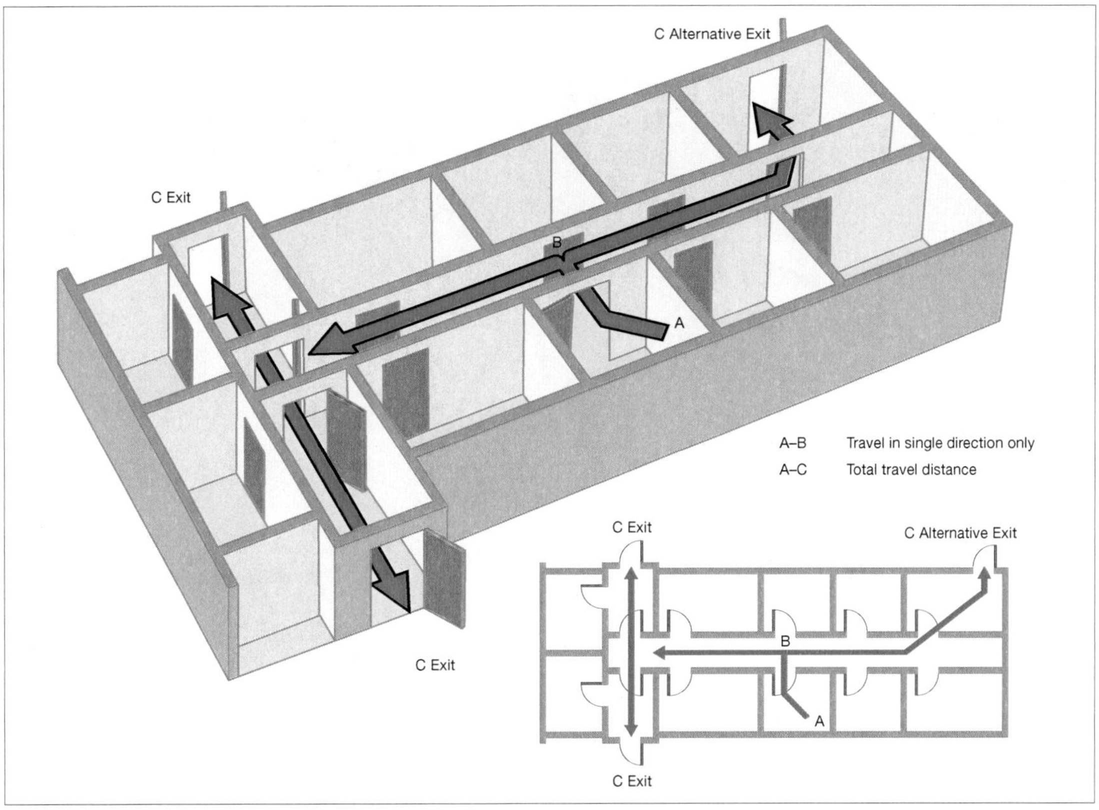

Basement, escape and protection

In all buildings with basements (other than very small basements), stairways serving upper floors should preferably not extend to the basement and in any case should not do so where they are the only stairway serving the upper floors. Any stairway that does extend from the basement to the upper floors should be seperated at basement level by a fire resisting lobby or corridor between the basement and the stairway. All basements used by more than 60 people or where there are no exits directly to a place of total safety, should have at least two protected escape stairways.

In high risk premises there should be an alternative stairway from the basement to ground level unless there is a suitable alternative route to the final exit.

Wherever possible all stairways to basements should be entered at ground level from the open air, and should be positioned so that smoke from any fire in the basement would not obstruct any exit serving the other floors of the building.

Where any stairway links a basement with the ground floor, the basement should be separated from the ground floor, preferably by two 30-minute fire doors, one at basement and one at ground floor level (see Figure 33).

Any floor over a basement should provide 60 minutes fire resistance. For smaller premises 30 minutes may be acceptable. Where this is impractical, and as long as no smoke can get through the floor, automatic smoke detection linked to a fire-alarm system which is audible throughout the premises could, as an alternative, be provided in the basement area. If in doubt, contact a competent person for more detailed advice.

Figure 33: Basement protection

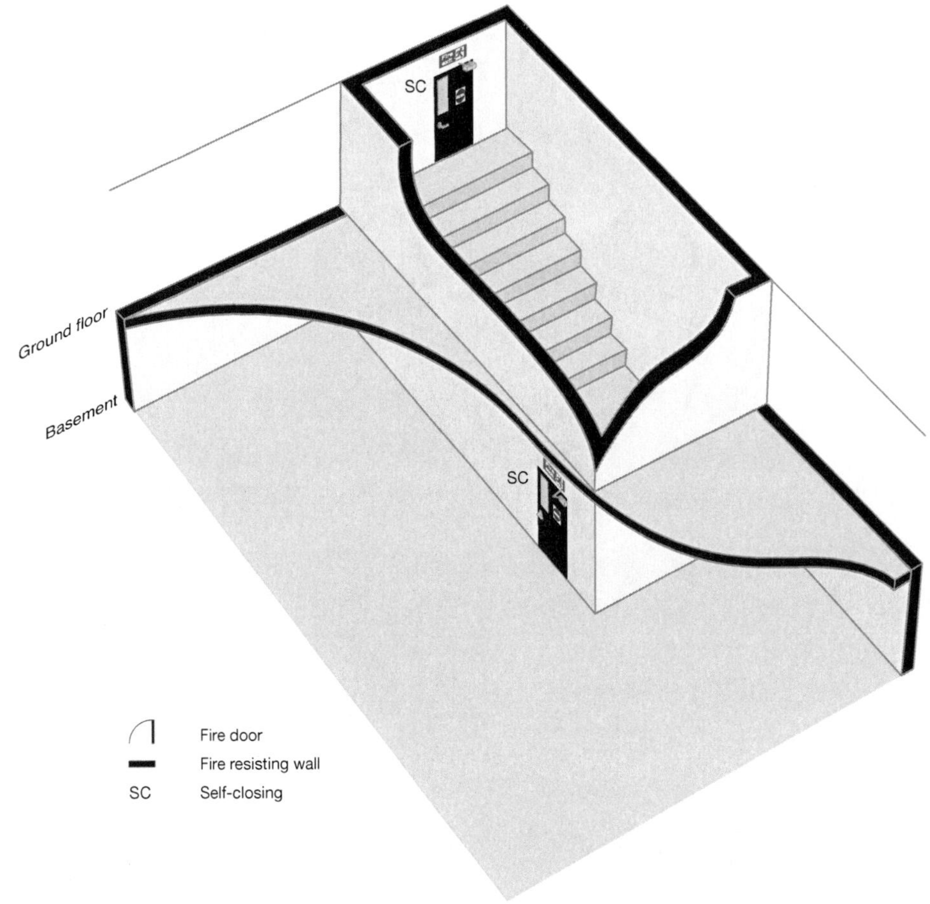

Subdivision of corridors

If your premises has corridors more than 30m long, then generally these corridors should be subdivided near the centre of the corridor with fire doors and, where necessary, fire-resisting construction to limit the spread of fire and smoke and to protect escape routes if there is a fire.

Where a corridor serves two exits from a floor, generally these corridors should be subdivided with fire doors to separate the two exits (see Figure 34).

Doors that are provided solely for the purpose of restricting the travel of smoke need not be fire doors, but will be suitable as long as they are of substantial construction, are capable of resisting the passage of smoke, and are self-closing. Smoke should not be able to bypass these doors, e.g. above a false ceiling, or via alternative doors from a room, or adjoining rooms, opening on either side of the subdivision.

Generally, false ceilings should be provided with barriers or smoke stopping over any fire doors. Where the false ceiling forms part of the fire-resisting construction this may not be necessary.

If you have doubts about subdivision of corridors, ask advice from a competent person.

Stairway enclosures

Stairways if unprotected from fire can rapidly become affected by heat and smoke cutting off the escape route and allowing fire to spread to other floors. However if adequately protected, escape stairways can be regarded as places of reasonable safety to enable people to escape to a place of total safety.

In most premises which are served by more than one stairway, it is probable that these stairways will be protected by fire-resisting construction and will lead to a final exit. If any floor has an occupancy of over 60 each storey should have at least two exits i.e. protected routes. The figure

Figure 34: Subdivision of corridor between two stairways or exits

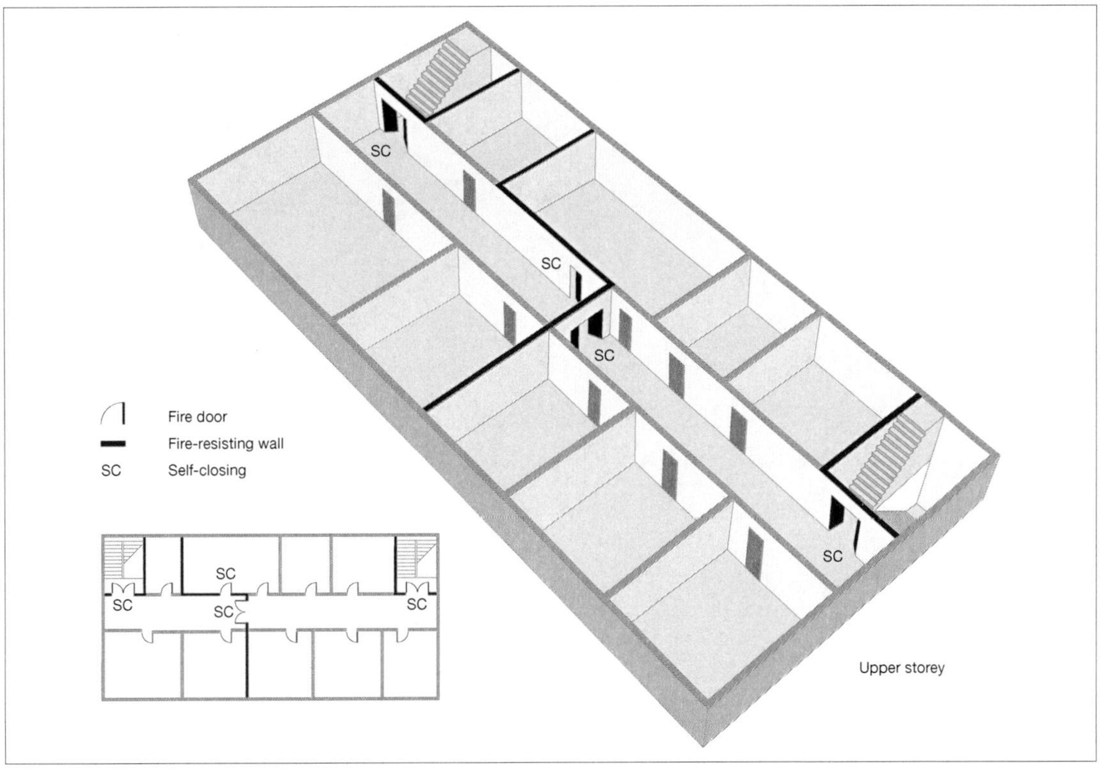

of 60 can be varied in proportion to the risk, lower risk slight increase, higher risk lower numbers of persons.

It is possible that you may have some stairways which have no fire protection to them. In this case they are not designed for escape and are normally known as accommodation stairways (see accommodation stairways on page 81).

If you have a protected stairway(s) then it is essential that you maintain that level of fire protection.

The benefit of protecting stairways from the effects of fire allows you to measure your travel distance from the furthest point on the relevant floor to the nearest storey exit rather than the final exit of the building.

If the building you occupy has floors which are occupied by different companies to your own you need to consider, as part of your fire risk assessment, the possibility that a fire may occur in another part of the building over which you may have no control and which may affect the protected stairway if allowed to develop unchecked. If your fire risk assessment shows that this may be the case and people using any floor would be unaware of a developing fire, then additional fire-protection measures may be required, e.g. an automatic fire-detection and warning system. If this is required you will need to consult and co-operate with other occupiers and building managers.

You may find that stairways in your building are provided with protected lobbies or corridors at each floor level (except the top floor) (Figure 35). Although these are not generally necessary for means of escape in multi-stairway buildings of less than 18m high, they may have been provided for other reasons (e.g. firefighting access). In all case protected corridors, lobbies and stairways must be kept clear of combustibles and obstructions.

In taller buildings (in general those over five storeys, excluding basements), the person who has overall control of common areas of the building may need to seek advice from a competent person.

Figure 35: Examples of a stairway with protected lobby/corridor approach

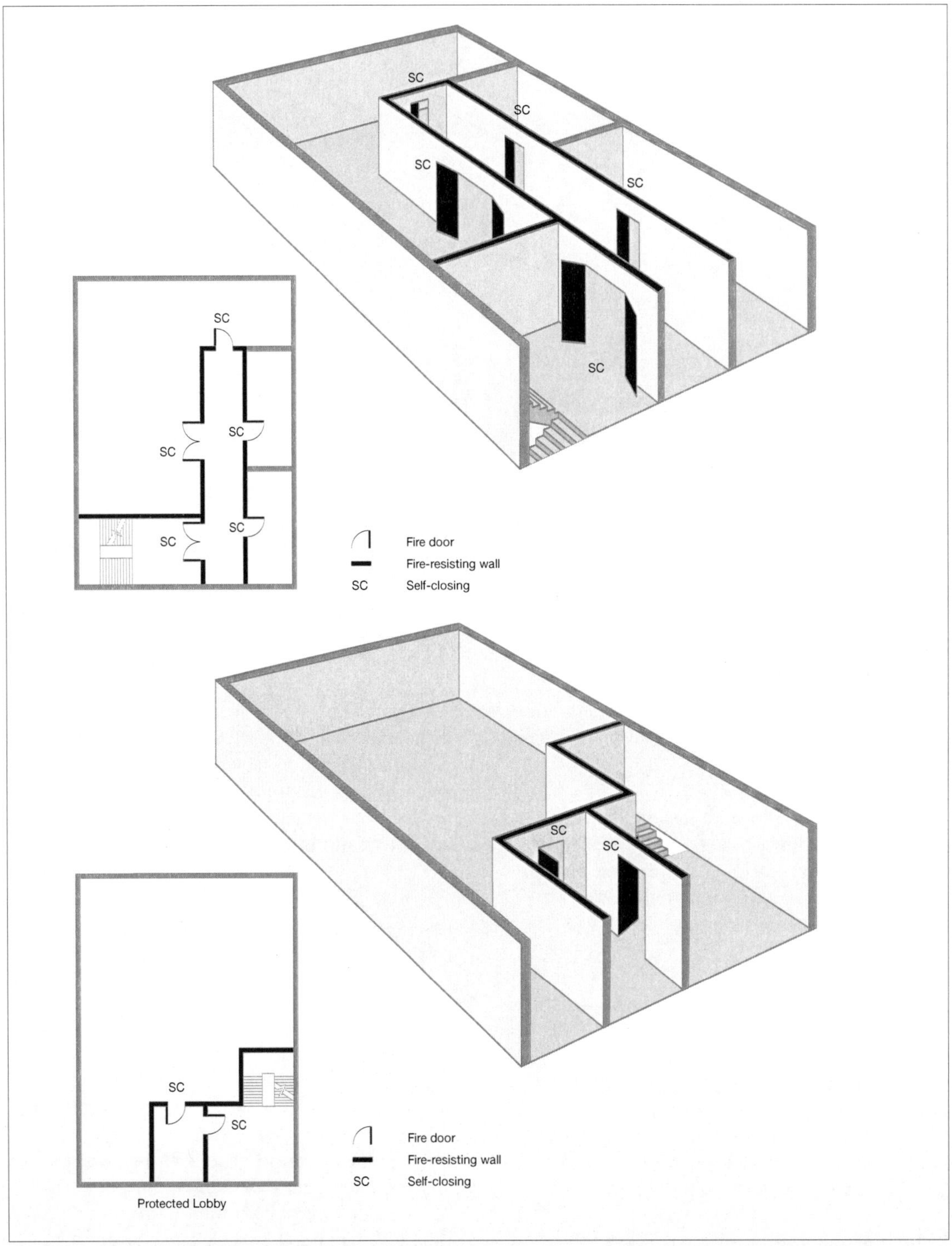

Ideally stairway enclosures should lead directly to a final exit. If your premises has only one stairway from the upper floor(s) which does not lead directly to a final exit, adopt one of the following arrangements:

- provide a protected route from the foot of the stairway enclosure leading to a final exit (Figure 36); or
- provide two exits from the stairway, each giving access to a final exit via routes which are separated from each other by fire-resisting construction (Figure 37).

Figure 36: Examples of a protected route from a stairway to a final exit

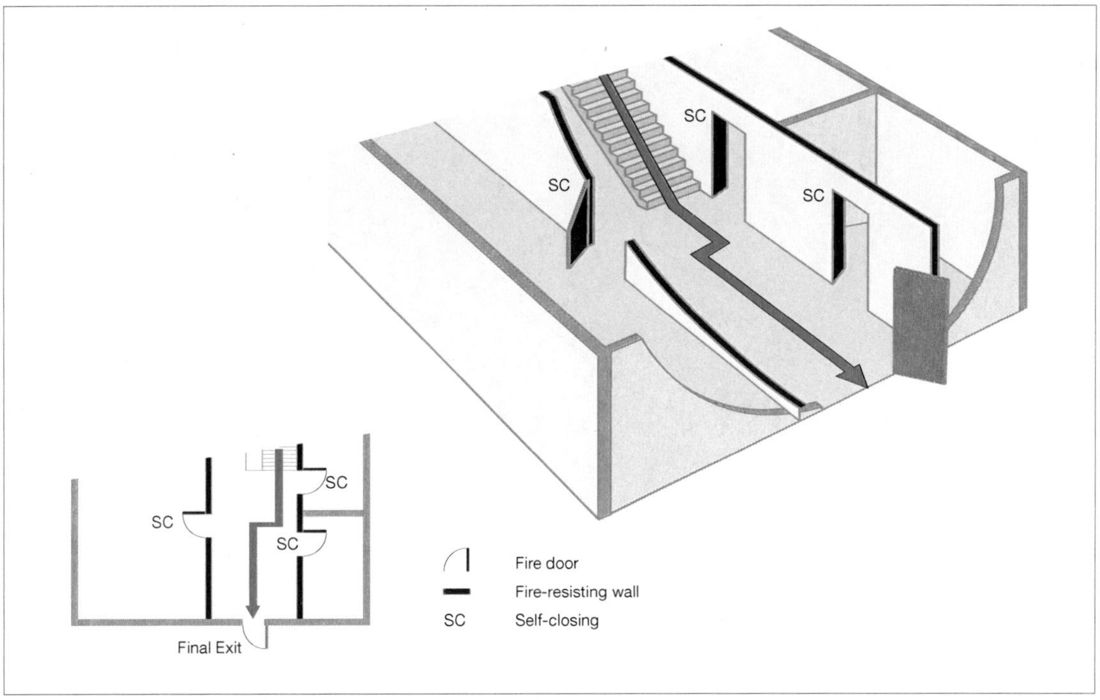

Figure 37: Examples of two escape routes from a stairway to final exits

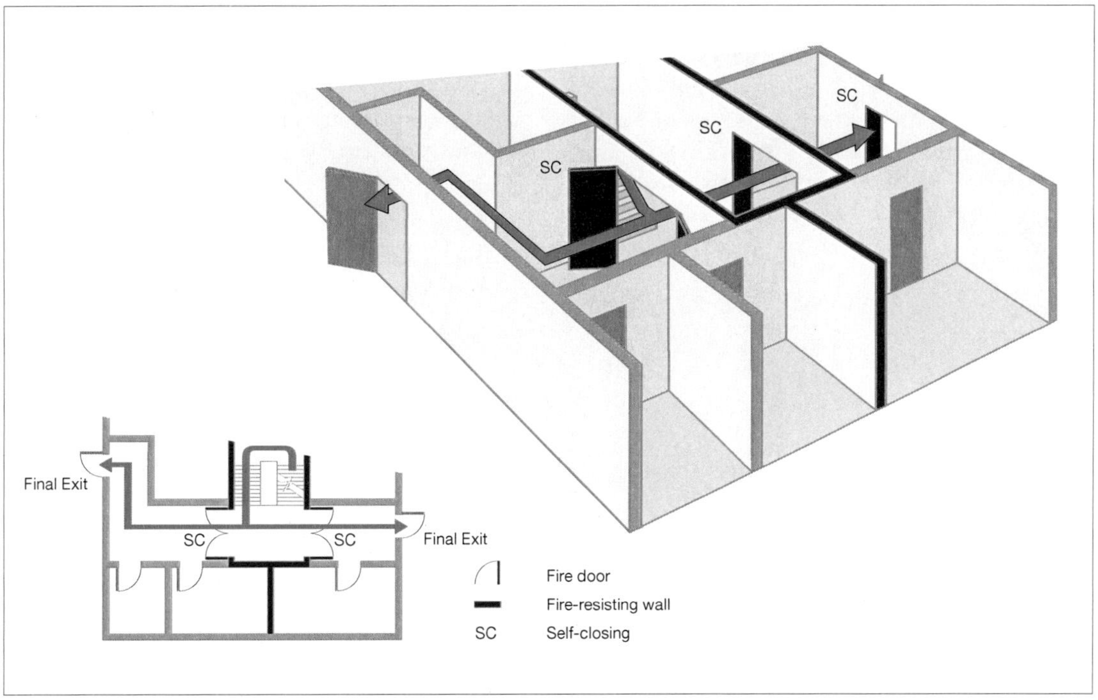

Figure 38: Separation of protected stairways

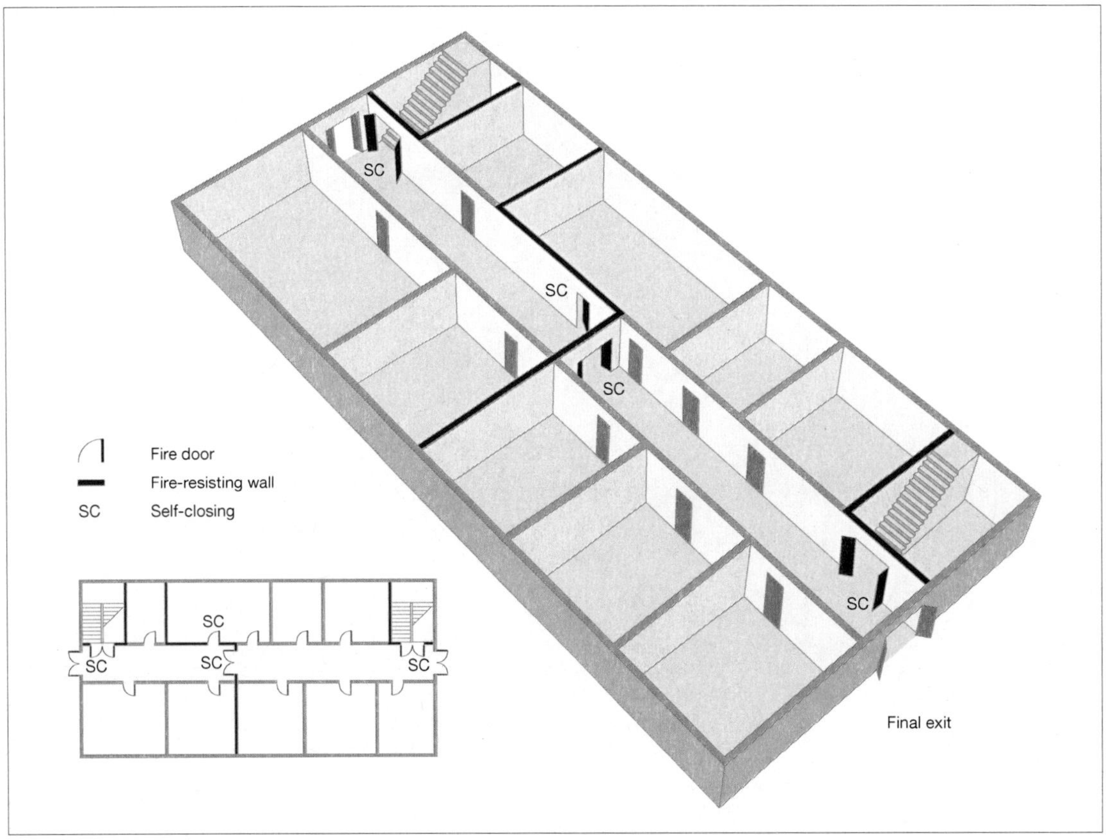

Separation of protected stairways

Where there are two or more protected stairways, the routes to final exits should be separated by fire-resisting construction so that fire cannot affect more than one escape route at the same time (see Figure 38).

Creating a stairway bypass route

No one should have to pass through a protected stairway to reach another stairway. Options to avoid this include:

- using intercommunicating doors between rooms adjacent to the stairway, such doors must be available at all times when the building is occupied (see Figure 39);
- using balconies and other features to bypass the stairway; or
- as long as there is enough space, create a bypass corridor around the stairway enclosure.

Figure 39: A stairway bypass route

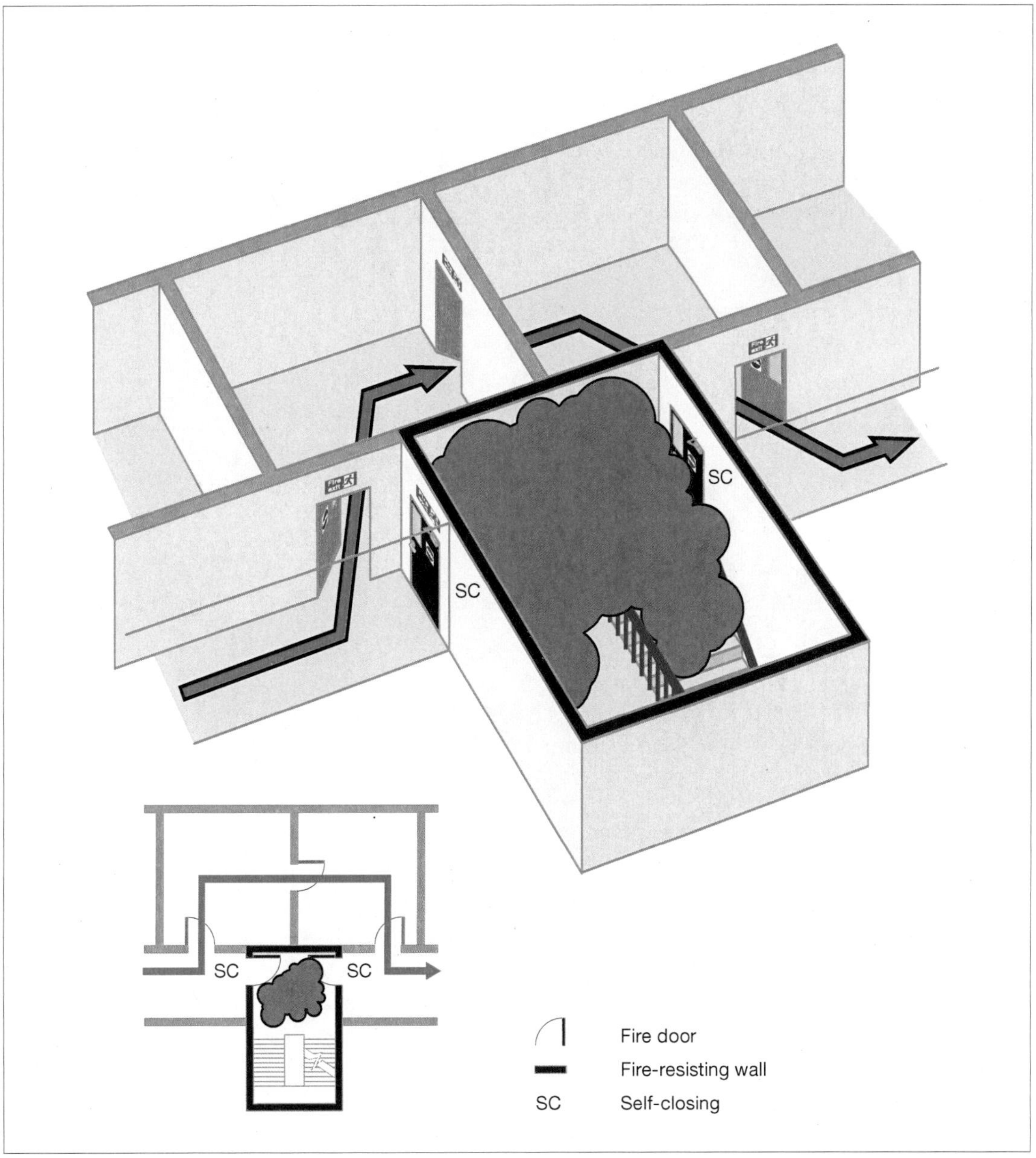

Reception areas

Reception or enquiry areas should only be located in protected stairways where the stairway is not the only one serving the upper floors, the reception area is small (less than 10m^2) and is of low fire risk.

Accommodation stairways

If you have stairways that are used for general communication and movement of people in the premises, and they are not designated as fire escape stairs then these are called 'accommodation stairways'. They may not require fire separation from the remainder of the floor as long as they do not pass through a compartment floor, or people do not have to pass the head of such a stairway in order to access a means of escape stairway. However, experience shows that many people will continue to use these as an escape route.

Accommodation stairways, escalators and travelators should not normally form an integral part of the calculated escape route capacity, however, where your fire risk assessment indicates that it is safe to do so, then you may consider them for that purpose. In these circumstances it may be necessary to seek advice from a competent person to verify this.

External stairways

To be considered a viable escape route, an external stairway should normally be protected from the effects of a fire along its full length. This means that any door, window (other than toilet windows) and walls within 1.8m horizontally and 9m vertically below any part of the stairway should be fire-resisting. Windows should be fixed shut and doors self-closing (Figure 40).

Figure 40: Protection to an external stairway

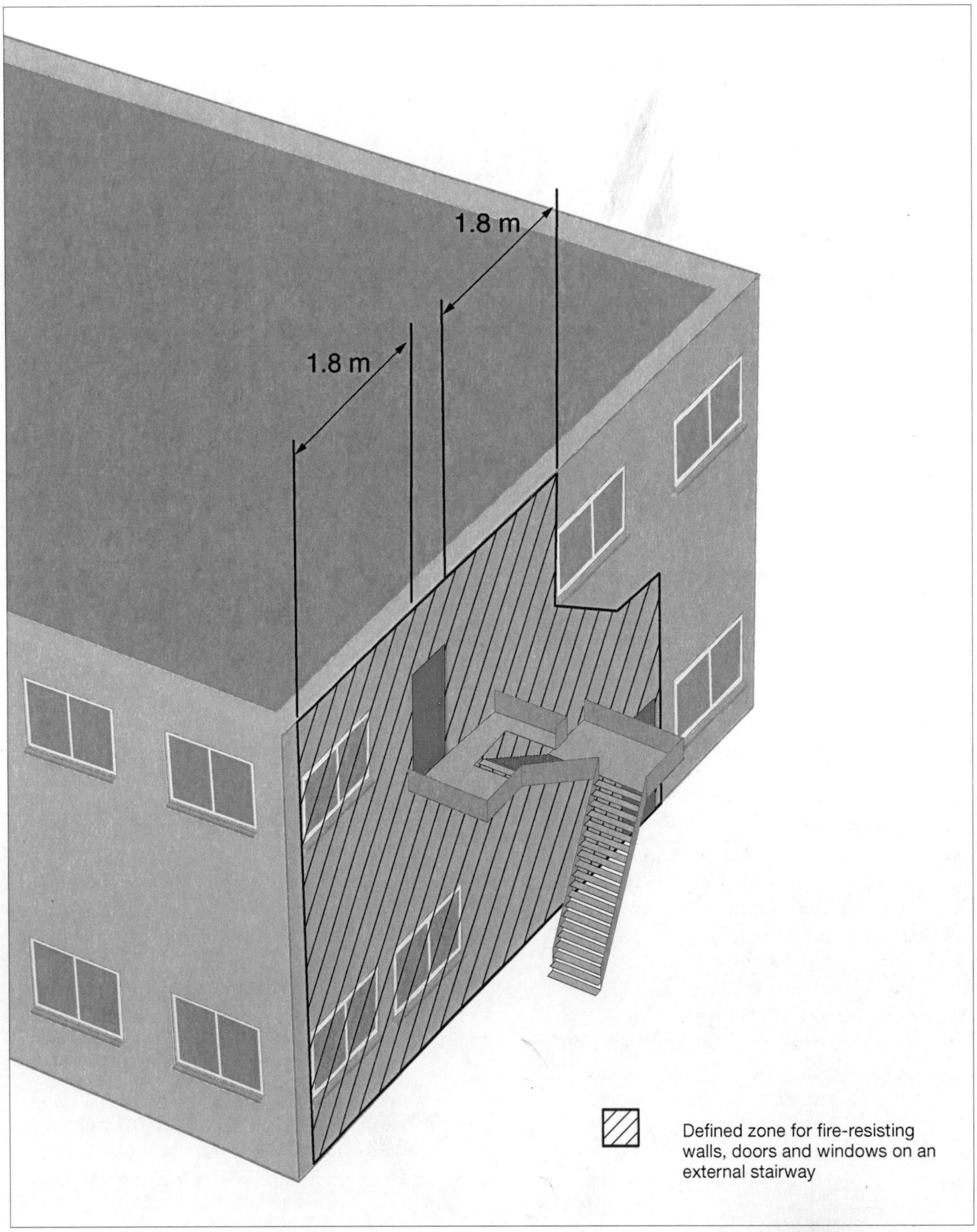

Monique frazer

These should not normally be used for members of the public, particularly where large numbers are likely to use them.

Consider protecting the external stairway from the weather as the treads may become slippery, e.g. due to algae, moss or ice. If this is not possible, you must ensure that the stairway is regularly maintained. Consider fixing non-slip material to the treads.

Spiral and helical stairways

Spiral and helical stairways are usually acceptable only in exceptional situations, e.g. for a maximum of 50 people who are not members of the public. The stairway should not be more than 9m in total height and not less than 1.5m in diameter with adequate headroom. A handrail should be continuous throughout the full length of the stairway.

However, spiral and helical stairways may be used as means of escape by more than 50 staff and may be used by the public if the stairways have been designed for the purpose. Further guidance is given in BS: 5395-2.[26] However, they are not usually suitable for young children.

Lifts

Due to the danger of the power supplies to a lift being affected by a fire, lifts not specifically designed as 'firefighting' or 'evacuation' lifts are not normally considered acceptable as a means of escape. However, where a lift and stairway for a means of escape are incorporated in a fire-resisting shaft which has a final exit from it at the access level and the lift has a separate electrical supply to that of the remainder of the building, than that lift subject to an agreed fire risk assessment, may be acceptable as a means of escape in case of fire.

Lifts are housed in vertical shafts that interconnect floors and compartments, therefore precautions have to be taken to protect people from the risk of fire and smoke spreading from floor to floor via the lift shaft. Such precautions may include:

- separating the lift from the remainder of the storey using fire-resisting construction and access via a fire door;
- ensuring the lift shaft is situated in a protected enclosure which may also be a stairway enclosure; and
- providing ventilation of at least $0.1m^2$ at the top of each lift-well to exhaust any smoke.

Roof exits

It may be reasonable for an escape route to cross a roof. Where this is the case, additional precautions will normally be necessary:

- The roof should be flat and the route across it should be adequately defined and well-illuminated where necessary with normal electric and emergency escape lighting. The route should be non-slip and guarded with a protective barrier.
- The escape route across the roof and its supporting structure should be constructed as a fire-resisting floor.
- Where there are no alternatives other than to use a roof exit, any doors, windows, roof lights and ducting within 3m of the escape route should be fire-resisting.
- Where an escape route passes through or across another person's property, you will need to have a robust legal agreement in place to allow its use at all times when people are on your premises.
- The exit from the roof should be in, or lead to, a place of reasonable safety where people can quickly move to a place of total safety.

A typical escape route across a roof is illustrated in Figure 41.

External escape routes should receive routine inspection and maintenance to ensure they remain fit for use. You will need to ensure that any legal agreements in place cover access for maintenance of the escape route.

Figure 41: An escape route across a roof

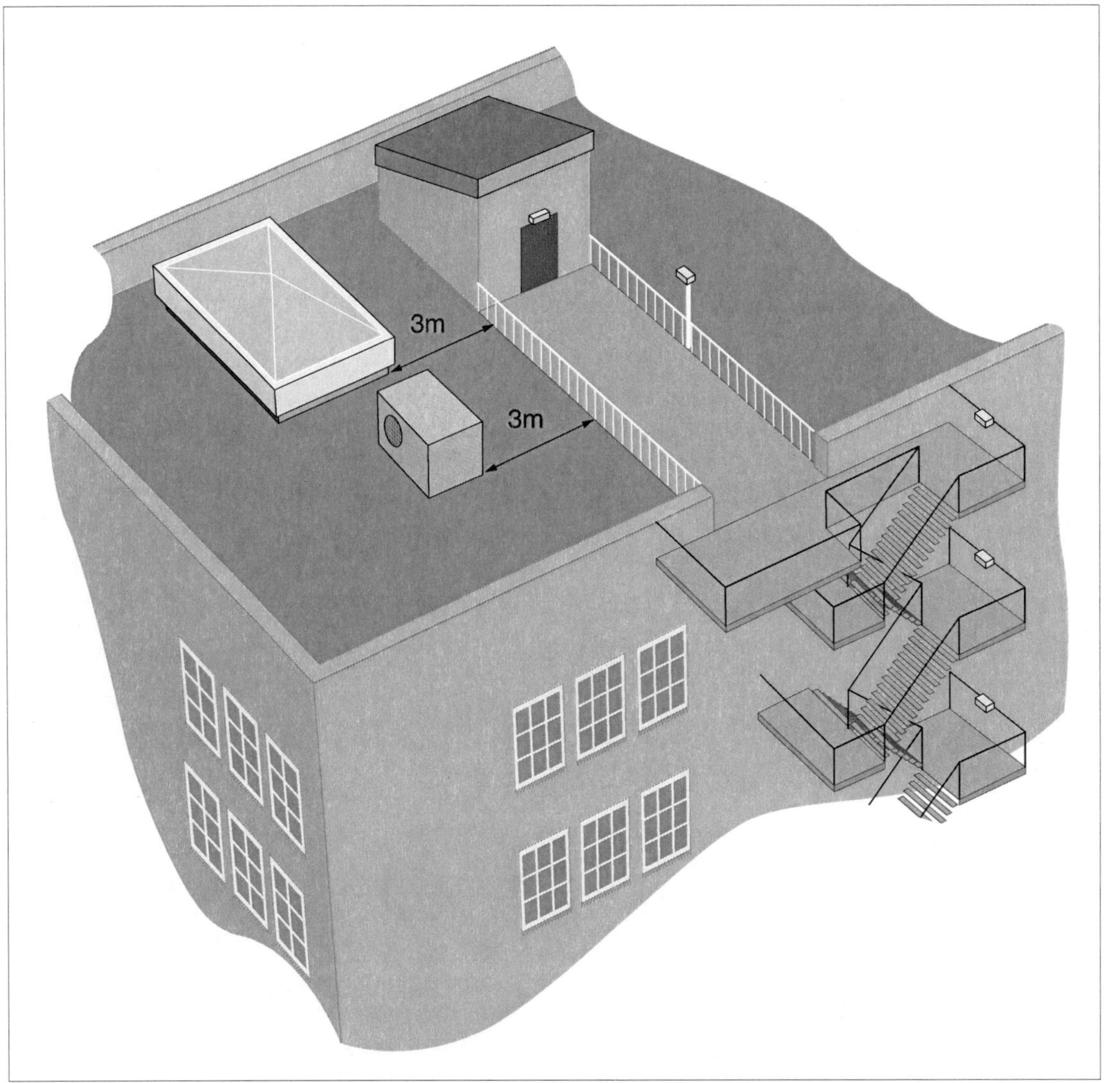

Revolving doors, wicket doors, sliding doors and roller shutters

Revolving doors should not normally be considered as escape doors unless the leaves fold outward to form a clear opening upon pressure from within or standard doors of the required exit width are provided next to the revolving door.

Ideally wicket doors or gates should have a minimum opening height of 1.5m. The bottom of the door should not be more than 250mm above the floor and the width should be preferably more than 500mm but not less than 450mm. Normally wicket doors will only be suitable for up to 15 members of staff however, in areas of a higher fire risk, this should be reduced to a maximum of three.

Loading and goods delivery doors, shutters (roller, folding or sliding), up-and-over doors and similar openings are not normally suitable for use as a final exit. However, they may be suitable for escape from areas of normal risk by small numbers of staff as long as they are not likely to be obstructed and can be easily and immediately opened manually, even if normally power-operated and the staff are familiar with the escape routes.

Sliding doors are not normally suitable on escape routes unless they are for the sole use of members of staff. Where provided, a notice with the words 'slide to open' with an arrow pointing in the direction of opening, should be permanently displayed at about eye level on the face of the door.

Final exit doors and escape away from the premises

Good escape routes to a final exit will be of little benefit if the occupants are not able to get out of the building and quickly disperse from the area to a place of total safety. It is also important to consider where people will go once they have evacuated from the premises.

The matters that you should consider include the following:

- Final exit doors should be quickly and easily openable without a key or code in the event of a fire. Where possible, there should be only one fastening. See Appendix B3 for more information on security fastenings.
- Final exit doors should not lead people into an enclosed area from which there is no further escape.
- Where a final exit discharges into an enclosed area, further access to a place of total safety should be available by means of further doors or gates that can be easily opened in a manner similar to the final exit.

4.2 Escape route layout

The examples listed in Table 3 on page 86 show typical escape route solutions for a range of common building layouts. In each case the solution is for a normal risk building unless otherwise illustrated.

These are not intended to be prescriptive or exhaustive but merely to help you understand how the principles of means of escape may be applied in practice.

They are illustrative of the key features of escape route layouts and not intended to be real building layouts or to scale.

You do not need to read all of this section, you only need to consider those figures and the accompanying text which most closely resemble your premises. If your premises do not resemble these then you should seek advice from a competent person. These examples are intended to represent your existing layout; they are not to be used as design guidance.

In all of these examples the following basic principles apply:

- The farthest point on any floor to the final exit or storey exit to a protected stairway is within the overall suggested travel distance (see Table 2).
- The route to and the area near the exit is kept clear of combustibles and obstructions.
- The fire-resisting stairway is kept clear of combustibles and obstructions.
- The escape route leads to a final exit.
- Where the stairway is not a protected stairway, the final exit is visible and accessible from the discharge point of the stairway at ground floor level.
- High-risk rooms do not generally open directly into a protected stairway.
- If your fire risk assessment shows that people using any floor would be unaware of a fire you may require additional fire-protection measures, e.g. an automatic fire-detection and warning system.
- There should be more than one escape route from all parts of the premises (rooms or storeys) except for areas or storeys with an occupancy of less than 60. The figure of 60 can be varied in proportion to the risk, for a lower risk there can be a slight increase, for a higher risk, lower numbers of persons should be allowed.

Table 3: Typical examples of escape route layouts

Single storey buildings (or the ground floor of a larger building)	
Ground floor premises with a single exit (including a mezzanine)	See Figure 42
Ground floor cellular layout	See Figure 43
Ground floor open plan layout	See Figure 44
Multi storey buildings with a single stairway	
Two-storey premises	See Figure 45
Two-storey premises, including basement	See Figure 46
Three-storey premises, including basement	See Figure 47
Larger three-storey premises, including basement	See Figure 48
Office premises, ground and up to three upper storeys	See Figure 49
Shop premises, ground and up to three upper storeys – protected lobbies/corridors	See Figure 50
Shop premises, ground and up to three upper storeys – automatic fire detection	See Figure 51
Multi-storey buildings with more than one stairway	
Example of a multi-storey building	See Figure 52
Tall building with a firefighting shaft	See Figure 53

If you do not have any of the stairway configurations given, and depending on the outcome of your fire risk assessment it may be that you can achieve an equivalent level of safety by other means.

The green arrows on the Figures 42–53 represent the travel distances given in Table 2 (page 68) which should be applied.

If your building has more than ground and three upper storeys, ask advice from a competent person.

Single-storey buildings (or the ground floor of a larger building)

Ground floor premises with a single exit (may include a mezzanine)

Part of your premises may have only a single exit. The example shown in Figure 42 will be generally acceptable provided that part of the premises served only by a single exit accommodates no more than 60 people in total.

If your fire risk assessment shows that people using mezzanine or cellular offices/stores would be unaware of fire elswhere in the building, it may require additional fire-protection measures, e.g. an automatic fire-detection and warning system.

Note: A mezzanine covering more than half of the floor area may need to be treated as a separate floor (see two-storey building).

Figure 42: Ground floor premises with a single exit (including a mezzanine)

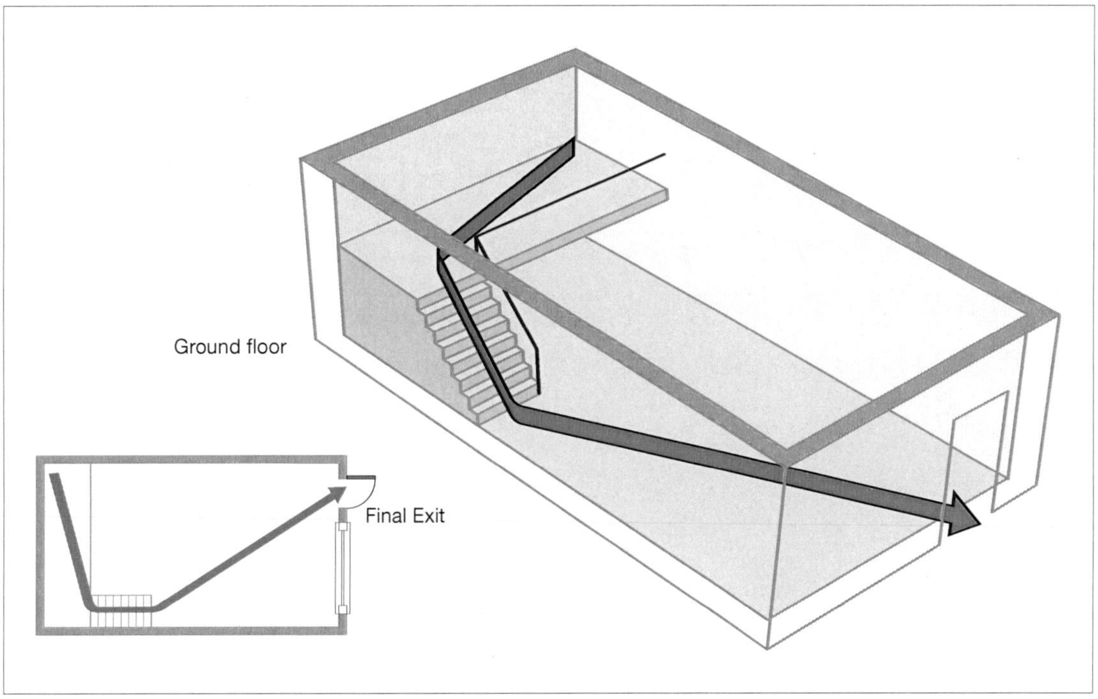

Ground floor premises with more than one exit

Figure 43 shows an acceptable example of a cellular layout for a ground floor premises with more than one exit.

Figure 44 shows an acceptable example of an open plan layout for a ground floor premises with more than one exit.

Figure 43: Ground floor cellular layout

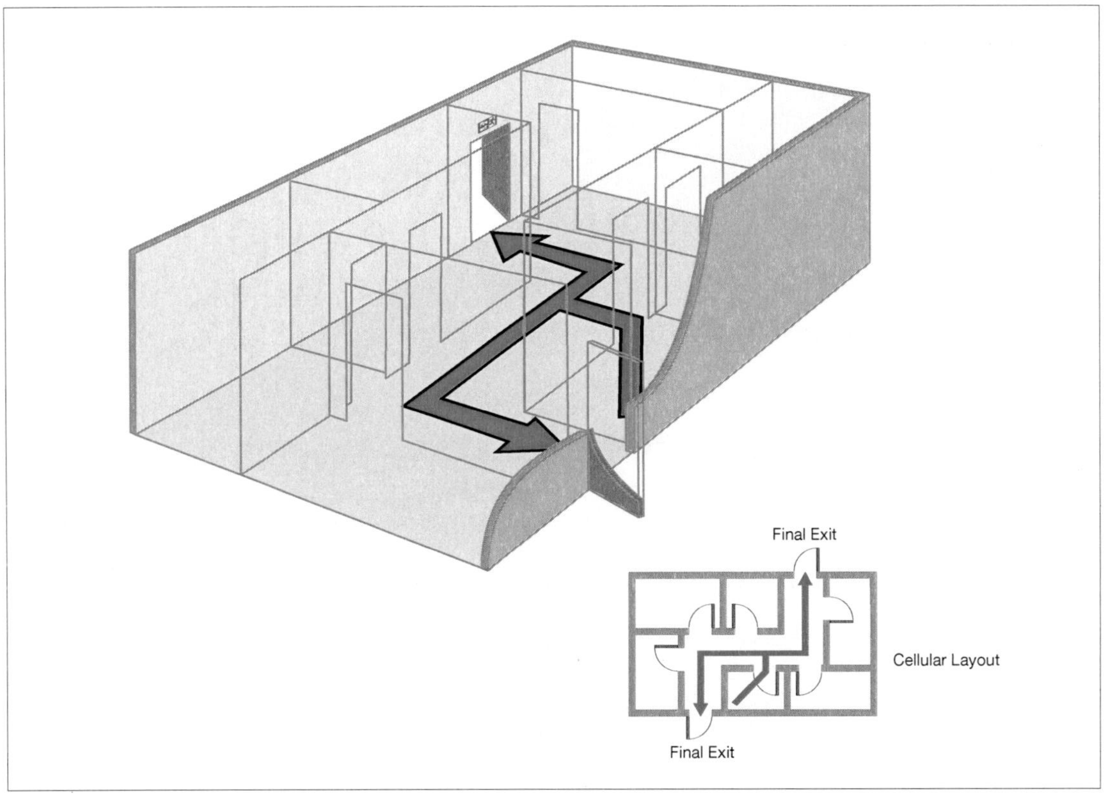

Figure 44: Ground floor open plan layout

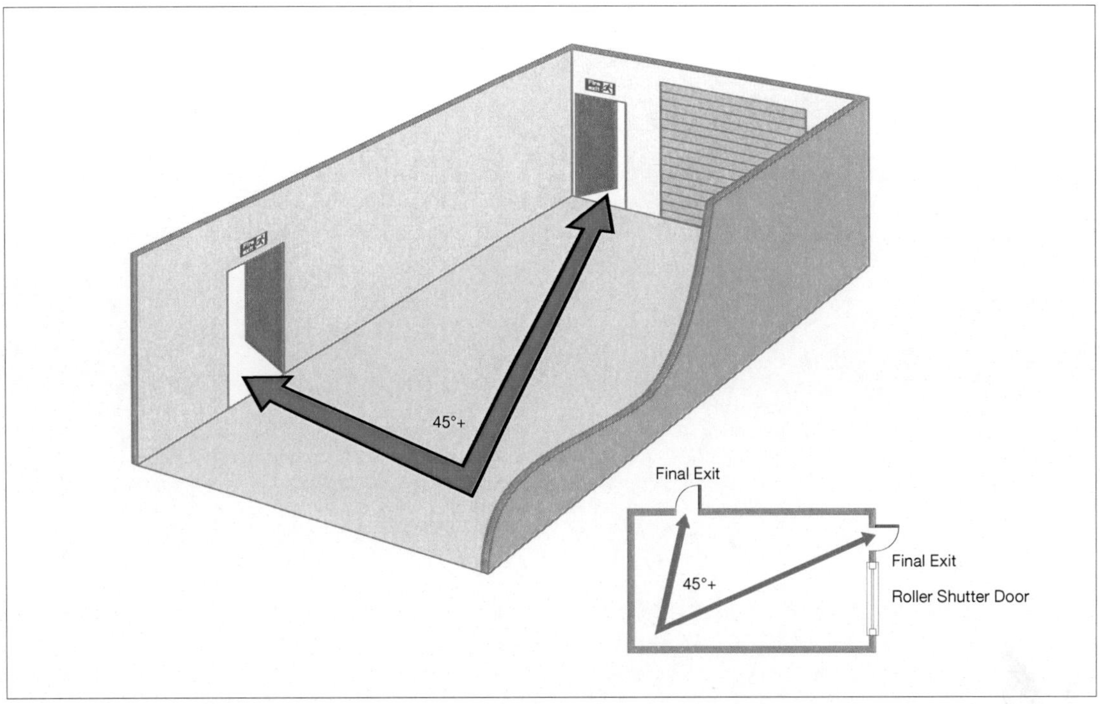

Multi-storey buildings with a single stairway

Two-storey premises

Figure 45 shows an acceptable layout for a two-storey premises with a single open stairway.

- The upper floor should accommodate no more than 60 people.
- The farthest point on each of the floors to the final exit is within the overall suggested travel distance for escape in one direction only (see Table 2, page 68).

Figure 45: Two-storey premises

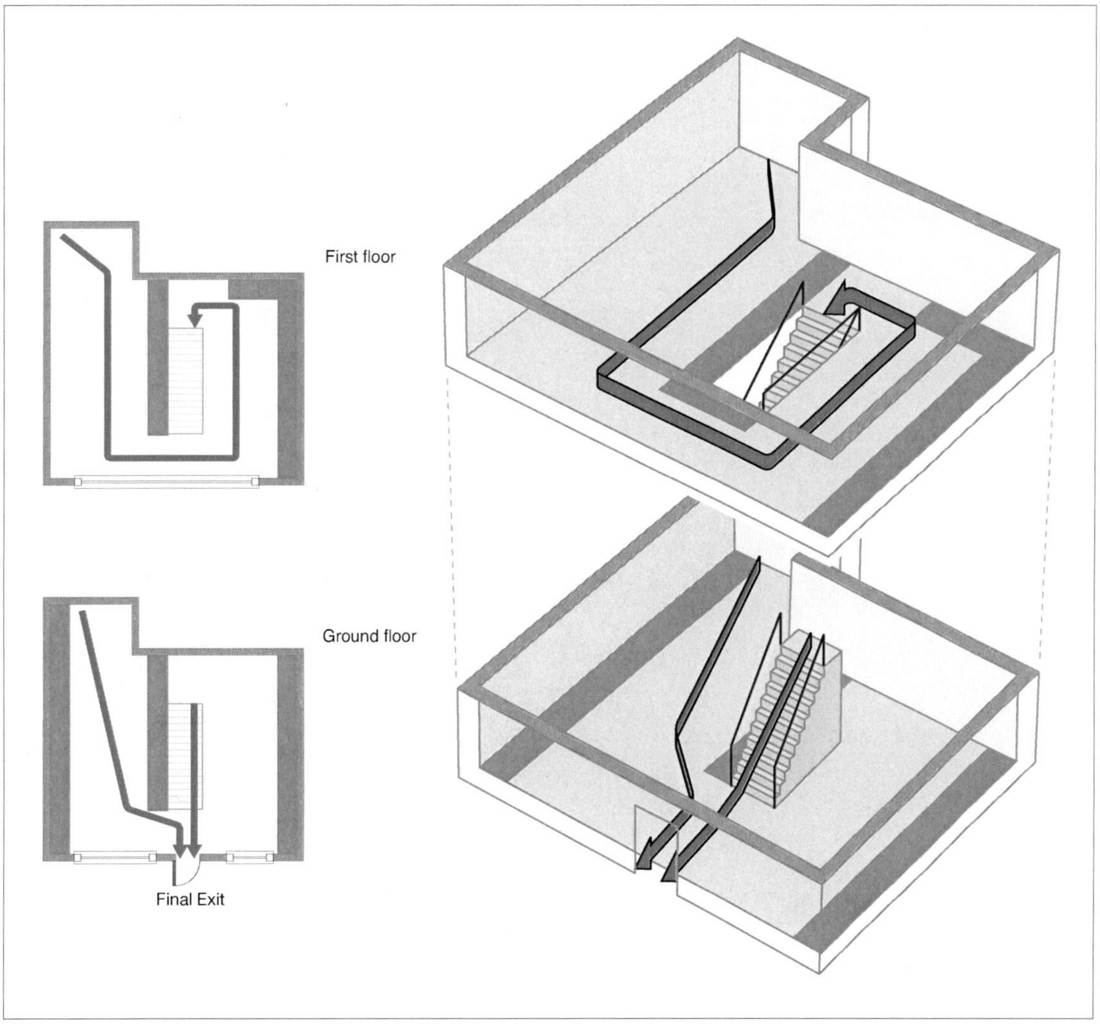

Two-storey premises, including basement
In premises with a basement and a simple layout, an open stairway is acceptable as indicated in Figure 46.

- The basement can accommodate no more than 60 people.
- The farthest point from any floor to the final exit is within the overall suggested travel distance (see Table 2, page 68).

Figure 46: Two-storey premises, including basement

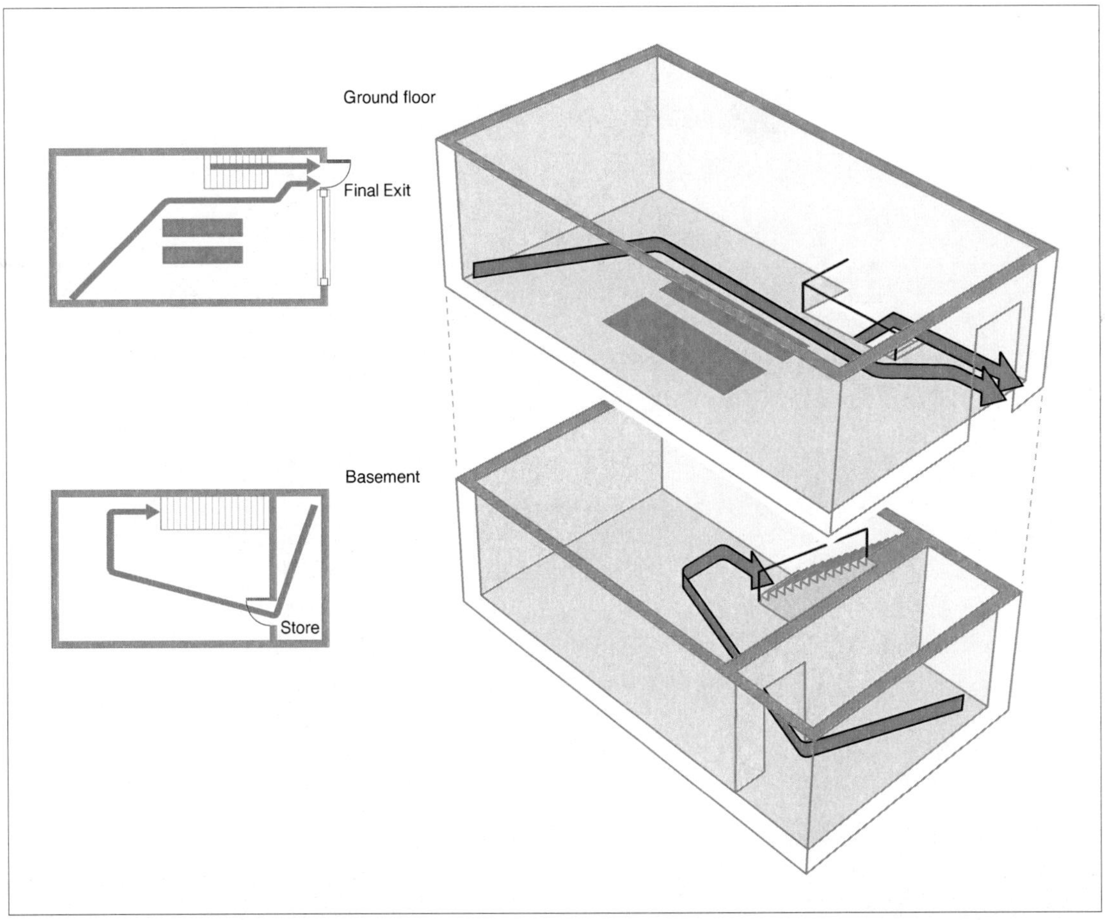

Three-storey premises, including basement
In premises with a basement, ground and first floor, served by individual stairways the layout in Figure 47 will be generally acceptable as long as the following apply:

- The basement should accommodate no more than 60 people.
- The first floor should accommodate no more than 60 people.
- The farthest point in the basement to the door to the stairway is within the overall suggested travel distance (see Table 2, page 68).
- The stairway from the basement to ground floor level is enclosed by fire-resisting construction and leads to a final exit.
- The farthest point on the first floor to the **final** exit is within the overall suggested travel distance.

Figure 47: Three-storey premises, including basement

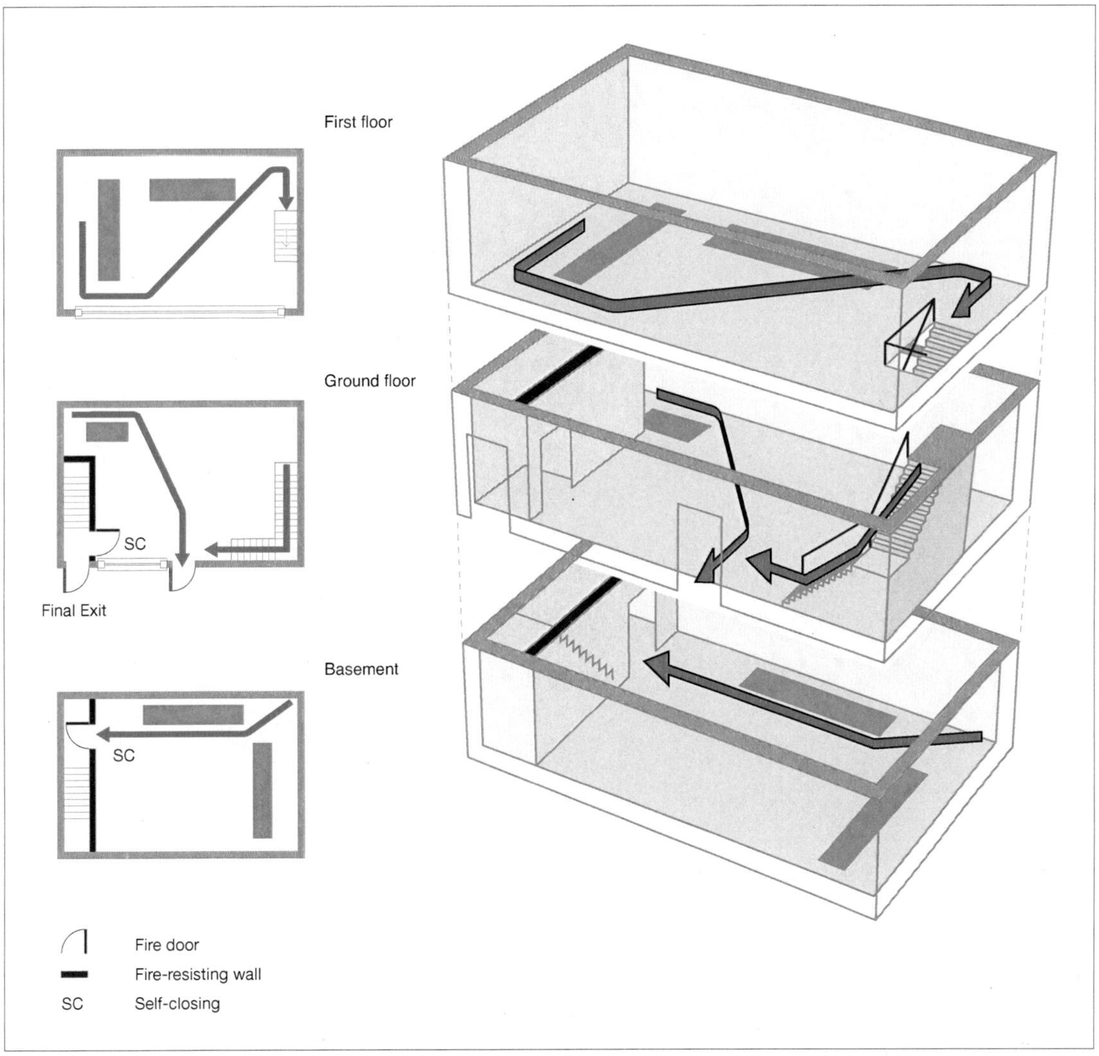

Larger three-storey premises, including basement
In larger premises with a basement, ground and first floor, served by a single stairway, the layout shown in Figure 48 will be generally acceptable as long as the following apply:

- The basement should accommodate no more than 60 people.
- The first floor should accommodate no more than 60 people.
- To overcome the restriction of travel distance the stairway has been completely enclosed in 30 minute fire-resisting construction and all doors onto the stairway are self-closing fire doors.
- The furthest point on all of the floors to the storey exit is within the overall suggested travel distance (see Table 2 page 68).
- Where the building incorporates a basement, any stairway from the basement is separated by a protected lobby or corridor between that basement and the protected stairway.

Figure 48: Larger three-storey premises, including basement

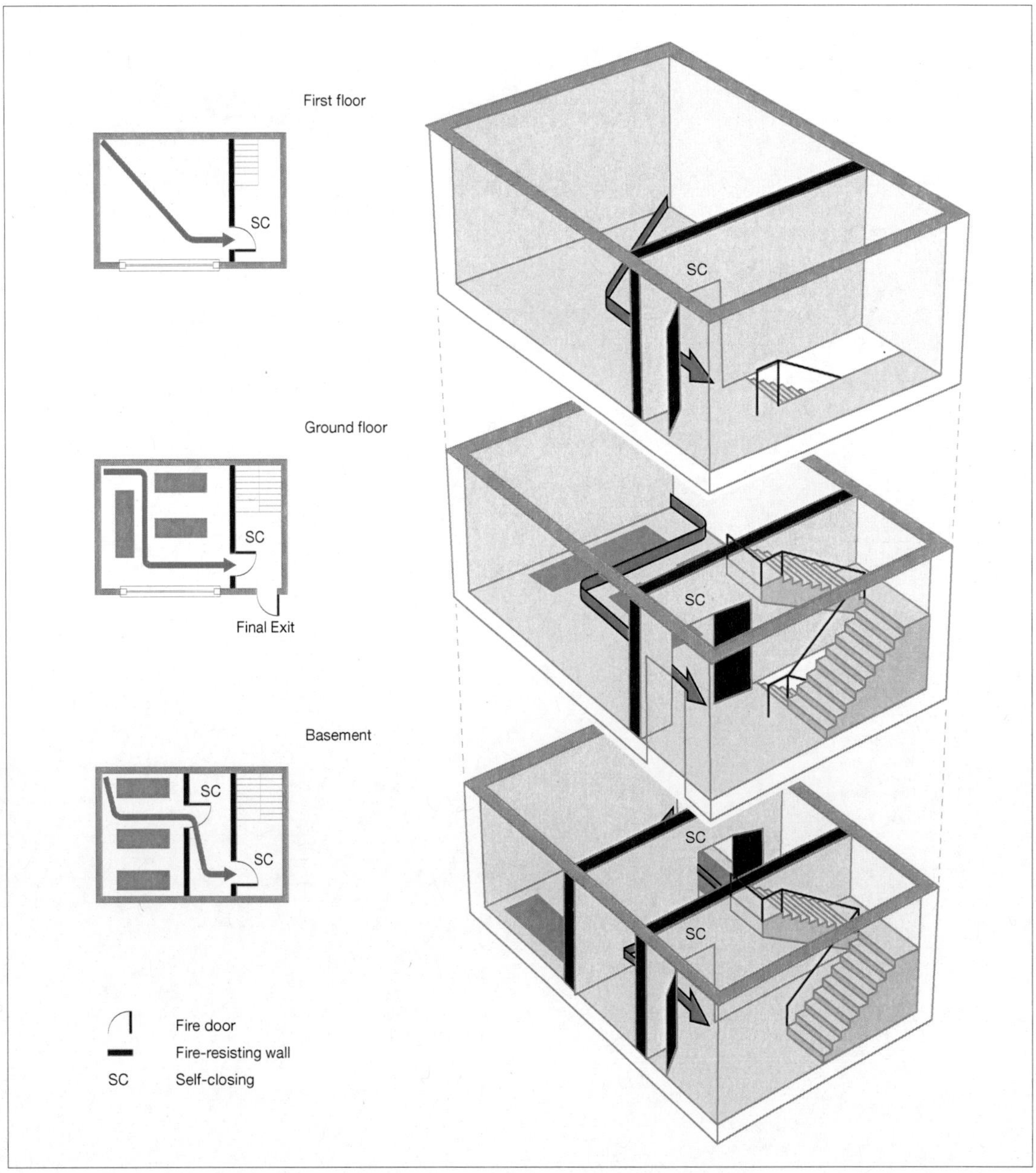

Office premises, ground and up to three upper storeys

In office premises only, with a ground floor and up to three upper storeys served by a single stairway, it is important to understand that you are unlikely to be able to meet the suggested travel distance to a final exit (see Table 2 page 68). In these circumstances it is necessary to protect the stairway by a fire-resisting enclosure as shown (Figure 49).

If the building you occupy has floors which are occupied by different companies to your own, you need to consider, as part of your fire risk assessment, the possibility that a fire may occur in another part of the building over which you may have no control and which may affect the protected stairway if allowed to develop unchecked. If your fire risk assessment shows that this may be the case and people using any floor would be unaware of a developing fire, you may require additional fire-protection measures, e.g. an automatic fire-detection and warning system. If this is required you will need to consult and co-operate with other occupiers and building managers.

The layout shown in Figure 49 will be generally acceptable as long as the following apply:

- The upper floors or any basement should each accommodate no more than 60 people.
- The furthest point on all of your floors to the storey exit is within the overall suggested travel distance (see Table 2 page 68).
- Where the building incorporates a basement, any stairway from the basement is separated by a fire-resisting lobby or corridor between that basement and the protected stairway.

Where your building has more than three upper storeys and a single stairway you should ask advice from a competent person.

Figure 49: Office premises, ground and up to three upper storeys

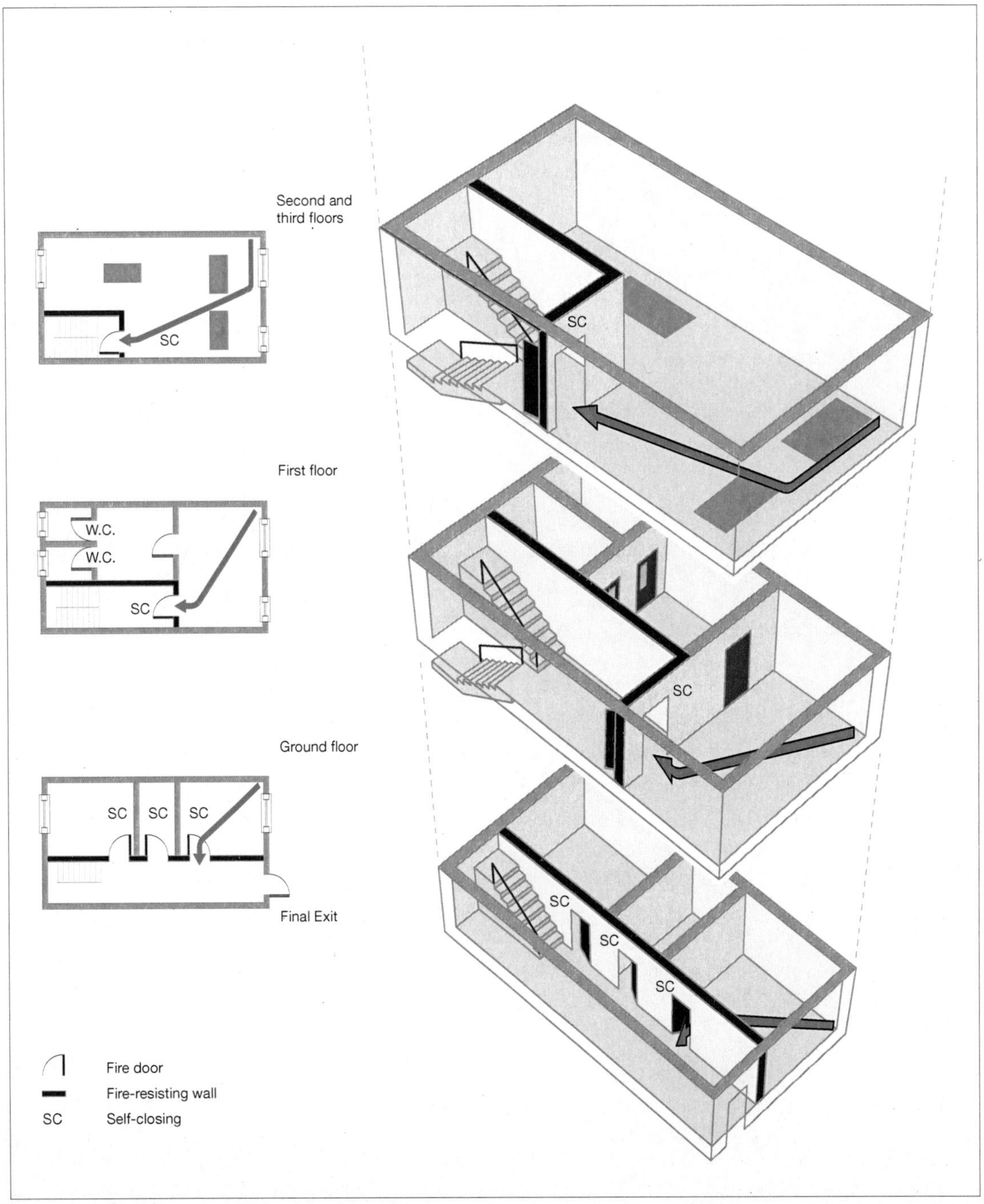

Shop premises, ground and up to three upper storeys

Because of the higher risk posed by shop premises, and to protect the escape route by preventing smoke from entering the stairway, a protected lobby or corridor approach between the stairway and all floors (other than the top floor) should be provided as shown in Figure 50.

The layout shown in Figure 50 will be generally acceptable as long as the following apply:

- The upper floors should each accommodate no more than 60 people.
- The furthest point on all of your floors to the lobbied storey exit is within the overall suggested travel distance (see Table 2 page 68).
- When a protected lobby or corridor approach to the stairway is employed the travel distance is measured to the storey exit and not the door to the lobby or corridor.

Alternatively automatic fire detection on all floors may be used instead of protected lobbies or corridors (see Figure 51), however, the stairway must still be protected.

If the building you occupy has floors which are occupied by different companies to your own, you need to consider, as part of your fire risk assessment, the possibility that a fire may occur in another part of the building over which you may have no control and which may affect the protected stairway if allowed to develop unchecked. If your fire risk assessment shows that this may be the case and people using any floor would be unaware of a developing fire, you may require additional fire-protection measures, e.g. an automatic fire-detection and warning system. If this is required you will need to consult and co-operate with other occupiers and building managers.

If your building has more than three upper storeys and a single stairway, ask advice from a competent person.

Figure 50: Shop premises, ground and up to three upper storeys – protected lobbies/corridors

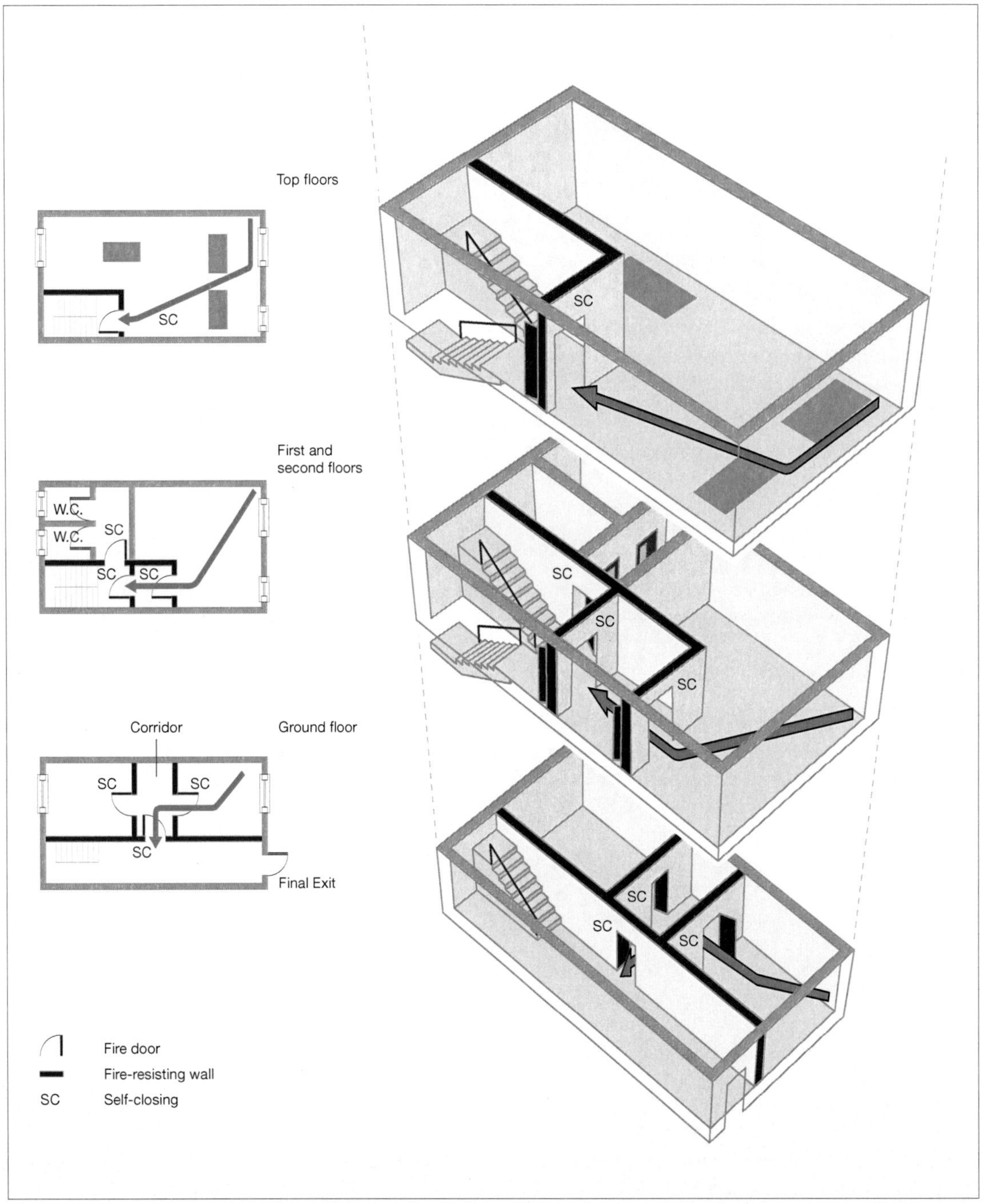

Figure 51: Shop premises, ground and up to three upper storeys – automatic fire detection

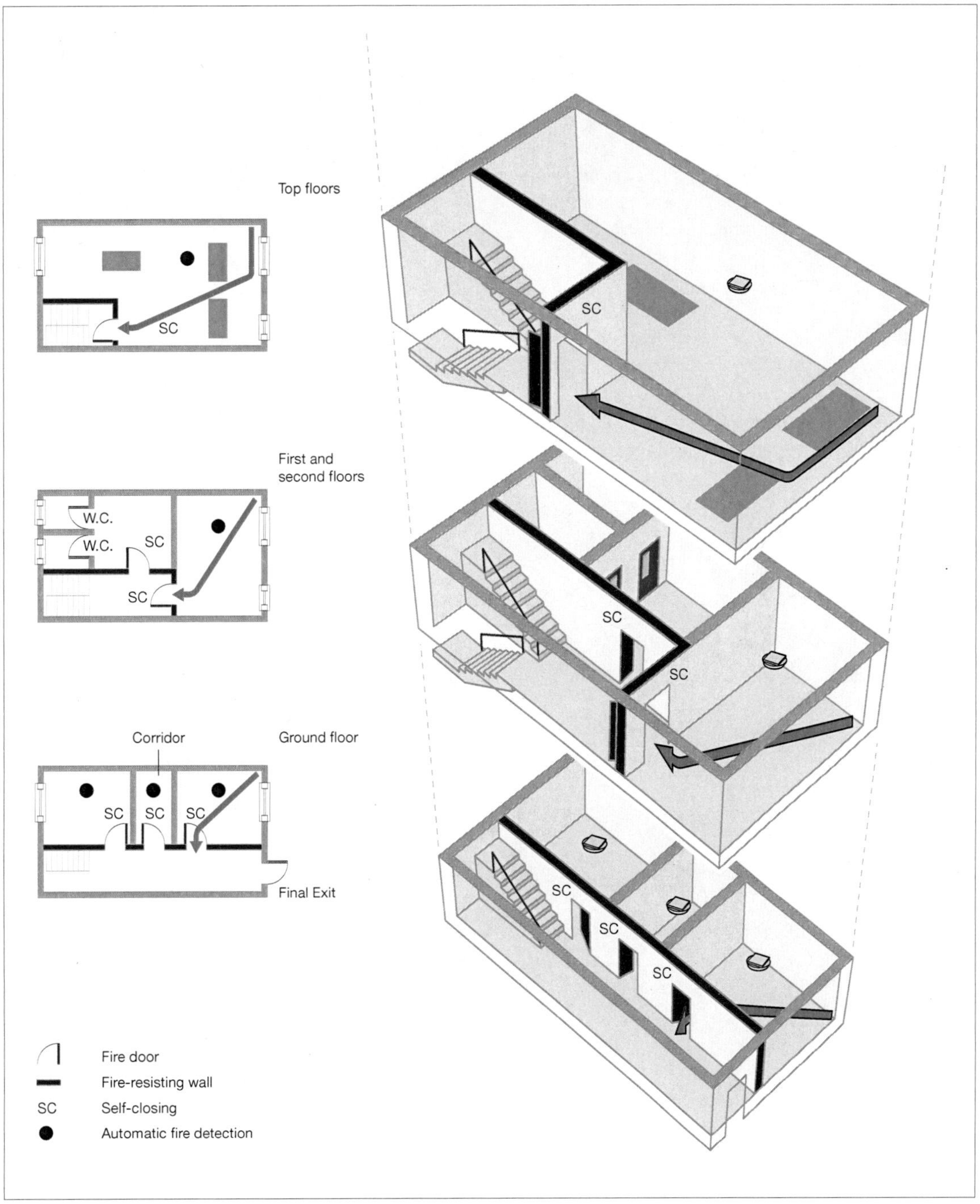

Multi-storey buildings with more than one stairway

Example of a multi-storey building

The layout shown in Figure 52 illustrates these principles. You also need to consider the following:

- Where the building incorporates a basement any stairway from the basement should be separated by a protected lobby or corridor between that basement and the protected stairway.
- If you do not have a protected stairway and, depending on the outcome of your fire risk assessment, it may be that you can achieve an equivalent level of safety by other means. However, it is strongly recommended that you ask advice from a competent person.
- These principles apply to taller buildings (with floors up to 18m high), however where your building has more than three upper storeys you should ask the advice of a competent person.

Figure 52: Example of a multi-storey building

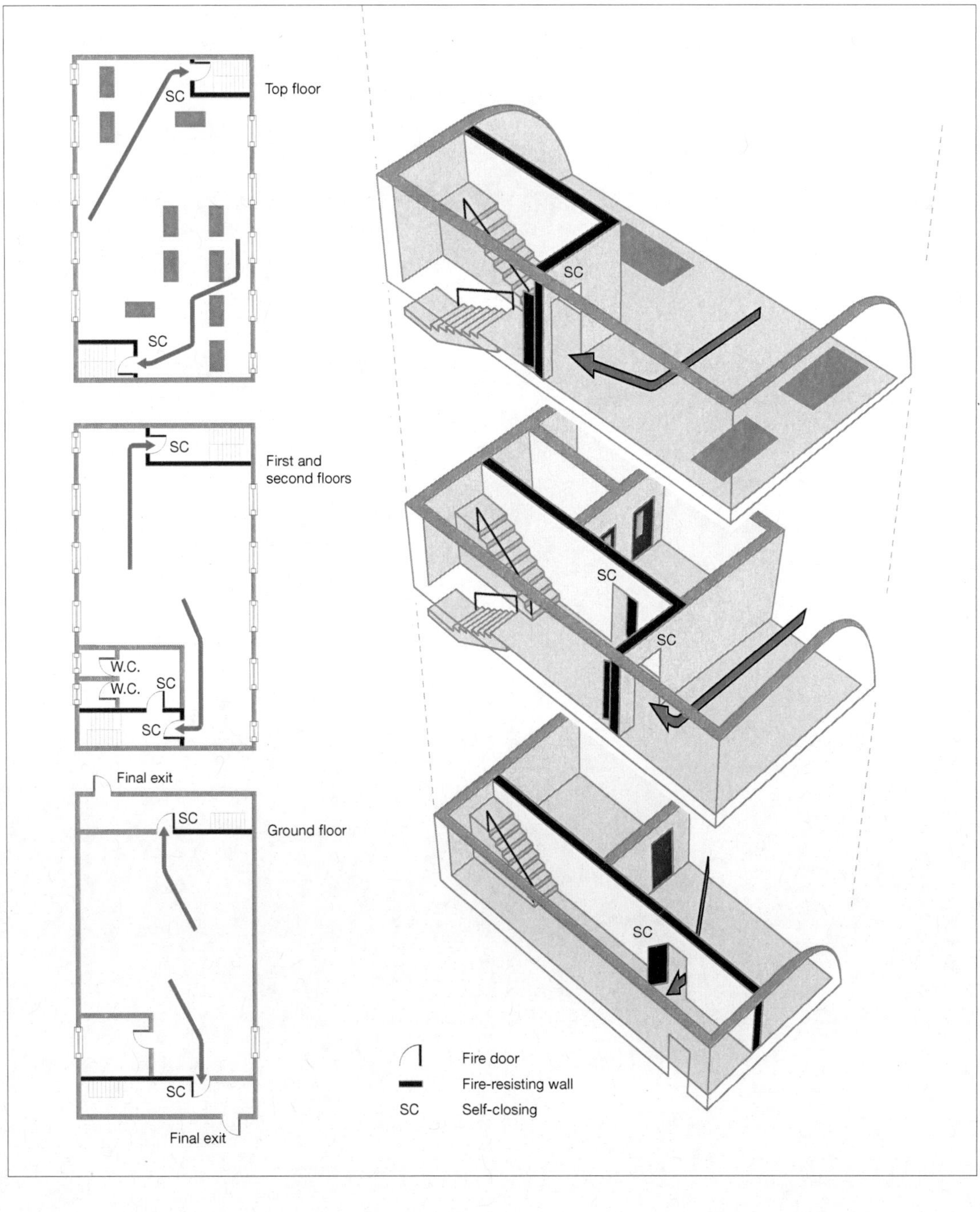

Tall building with a firefighting shaft
Figure 53 shows a multi-storey building fitted with a firefighting shaft which is required for specific types of buildings including multi-stairway buildings with a floor higher than 18m.

If the premises you occupy are situated in a building like this, you should ask the advice of a competent person.

Figure 53: Tall building with a firefighting shaft

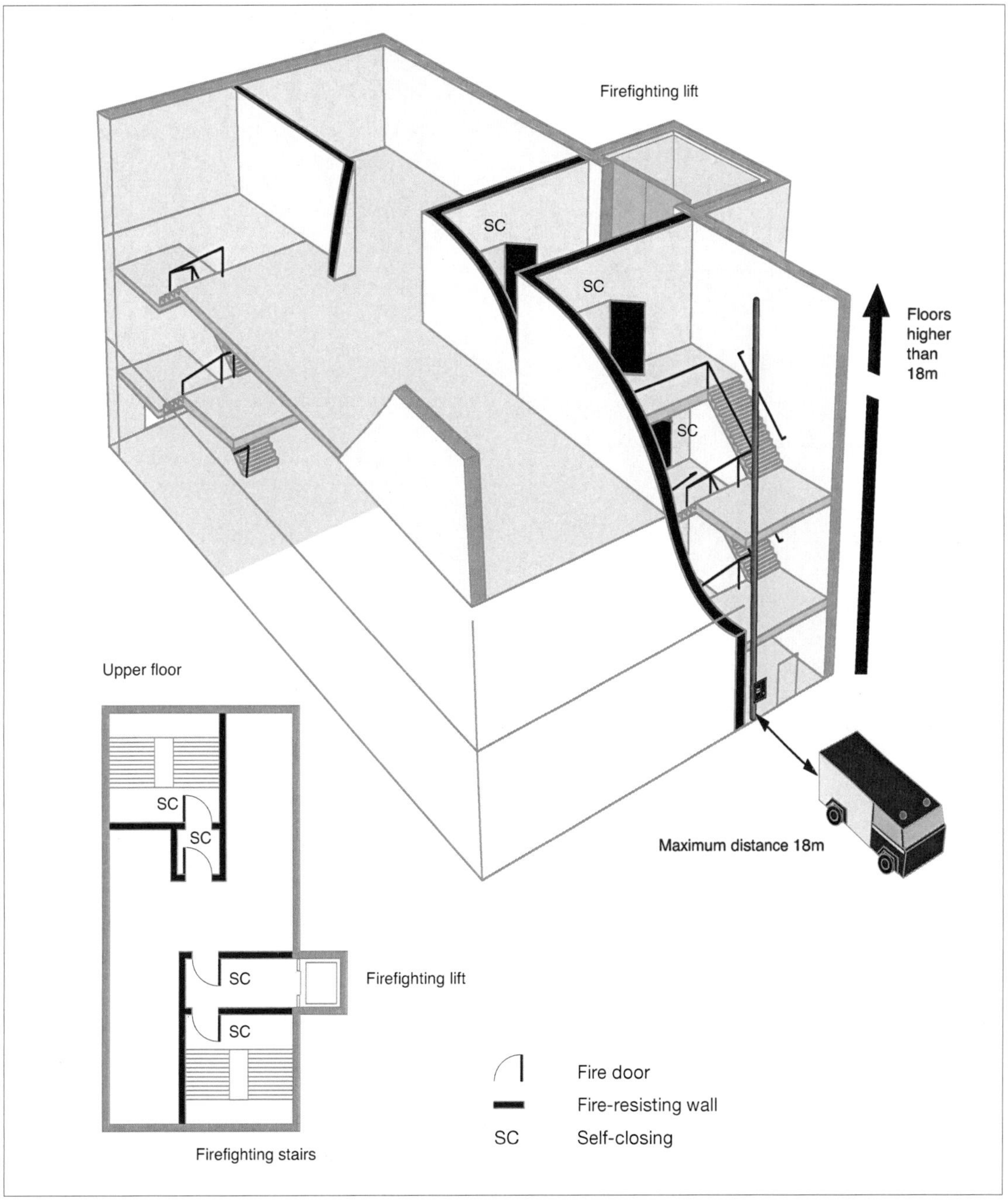

Section 5 Further guidance on emergency escape lighting

The primary purpose of emergency escape lighting is to illuminate escape routes but it also illuminates other safety equipment.

The size and type of your premises and the risk to the occupants will determine the complexity of the emergency escape lighting required. Borrowed lighting may be suitable in small premises where the light is from a dependable source, e.g. street lamps, and it will adequately illuminate escape routes. Where borrowed lighting is not suitable, then a number of torches, in strategic positions, can be considered.

Single 'stand-alone' escape lighting units may be sufficient in small premises and these can sometimes be combined with exit or directional signs (Figure 54). The level of general illumination should not be significantly reduced by the sign.

In larger more complex premises a more comprehensive system of fixed automatic escape lighting is likely to be needed. This will be particularly true in premises with extensive basements or where there are significant numbers of staff or members of the public.

You will have identified the escape routes when carrying out your fire risk assessment and need to ensure that they are all adequately lit. If there are escape routes that are not permanently illuminated by normal lighting, such as external stairs, then a switch, clearly marked 'Escape lighting', or some other means of switching on the lighting should be provided at the entry to that area/stairs.

An emergency escape lighting system should normally cover the following:

- each exit door;
- escape routes;
- intersections of corridors;
- outside each final exit and on external escape routes;
- emergency escape signs;
- stairways so that each flight receives adequate light;
- changes in floor level;
- windowless rooms and toilet accommodation exceeding $8m^2$;
- firefighting equipment;
- fire alarm call points;
- equipment that would need to be shut down in an emergency;
- lifts; and
- areas in premises greater than $60m^2$.

It is not necessary to provide individual lights (luminaires) for each item above, but there should be a sufficient overall level of light to allow them to be visible and usable.

Emergency escape lighting can be both 'maintained', i.e. on all the time, or 'non-maintained' which only operates when the normal lighting fails. Systems or individual lighting units (luminaires) are designed to operate for durations of between one and three hours. In practice, the three-hour design is the most popular and can help with maintaining limited continued use of the premises during a power failure (other than in an emergency situation).

Figure 54: Luminaires

Emergency escape lighting (luminaires) can be stand-alone dedicated units or incorporated into normal light fittings. There are highly decorative versions of these for those areas that demand aesthetically pleasing fixtures. Power supplies can be rechargeable batteries integral to each unit, a central battery bank or an automatic start generator.

To complement emergency escape lighting, people, especially those unfamiliar with the premises, can be helped to identify exit routes by the use of way-guidance equipment. Way-guidance systems usually comprise photo-luminescent material, lines of LEDs, or strips of miniature incandescent lamps, forming a continuous marked escape route at lower level (Figure 55). These systems have proved particularly effective when people have to escape through smoke, and for partially-sighted people. They can be particularly useful in premises where they can provide marked routes on floors and in multi-storey premises they can direct people to escape routes which are seldom used.

If you decide that you need to install emergency escape lighting or to modify your existing system, any work should be carried out by a competent person in accordance with the appropriate standards. Further guidance is given in BS 5266-1[28] and BS 5266-8.[27]

Figure 55: A 'way-guidance' system

Maintenance and testing of emergency escape lighting

All emergency escape lighting systems should be regularly tested and properly maintained to an appropriate standard. Most existing systems will need to be manually tested. However, some modern systems have self-testing facilities that reduce routine checks to a minimum.

Depending on your type of installation you should be able to carry out most of the routine tests yourself. The test method will vary. If you are not sure how to carry out these tests you should contact your supplier or other competent person.

Figure 56: A test key

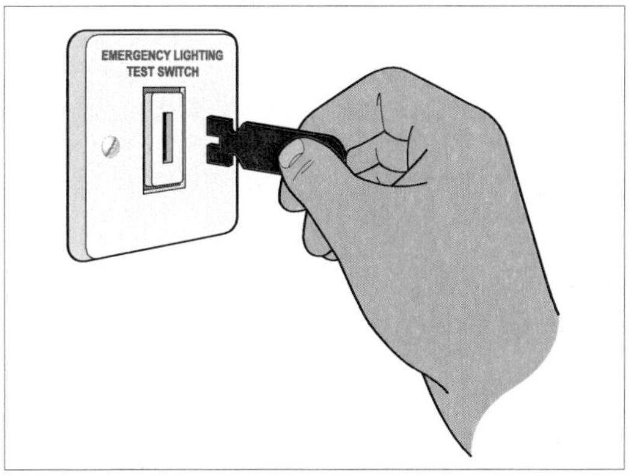

Test facilities often take the form of a 'fishtail' (Figure 56) key inserted in a special switch either near the main fuse board or adjacent to relevant light switches.

Typically, testing would include:

- a daily visual check of any central controls;
- a monthly function test by operating the test facility for a period sufficient to ensure that each emergency lamp illuminates; and
- an annual full discharge test.

Particular care needs to be taken following a full discharge test. Batteries typically take 24 hours to re-charge and the premises should not be re-occupied until the emergency lighting system is fully functioning unless alternative arrangements have been made. See BS 5266-8[27] for more information.

It is good practice to keep a record of tests.

Section 6 Further guidance on signs and notices

Escape signs

In simple premises, a few signs indicating the alternative exit(s) might be all that is needed. In larger and more complex premises, a series of signs directing people along the escape routes towards the final exit might be needed.

Many people with poor vision retain some sight and are able to recognise changing or contrasting colour to provide them with visual clues when moving around a building.* It may sufficient to paint any columns and walls in a contrasting colour and to highlight changes in level by, for example, making the nosing to step and stair treads a contrasting colour (see BS 8300[14]).

For people with no sight, a well-managed 'buddy system', continuous handrails or a sound localisation system (which helps people to move towards an alert sound), or the installation of more tactile aids may be appropriate.

Exit signs should be clearly visible whenever the public, staff and contractors are present.

Positioning of escape signs

The presence of other signs in shops (such as advertising and customer information) or shop dressing, can distract attention from, or obscure the visibility of escape signs. This could affect people's ability to see and understand escape signs, particularly if there is a fire evacuation. Always ensure that escape signs are not overwhelmed.

Escape signs should meet the following criteria:

- They should provide clear, unambiguous information to enable people to safely leave a building in an emergency.
- Every escape route sign should, where necessary, incorporate, or be accompanied by, a directional arrow. Arrows should not be used on their own.
- If the escape route to the nearest exit is not obvious then it should be indicated by a sign(s).
- Signs should be positioned so that a person escaping will always have the next escape route sign in sight.
- Escape signs should be fixed above the door in the direction of escape and not be fixed to doors, as they will not be visible if the door is open.
- Signs mounted above doors should be at a height of between 2.0m and 2.5m above the floor.
- Signs on walls should be mounted between 1.7m and 2.0m above the floor.
- Mounting heights greater than 2.5m may be used for hanging signs, e.g. in large open spaces or for operational reasons, but care should be taken to ensure that such signs are both conspicuous and legible. In such case larger signs may be necessary.
- Signs should be sited at the same height throughout the escape route, so far as is reasonably practicable.

Escape sign design

For a sign to comply with safety signs and signals regulations it must be pictographic (see Figures 57 and 58). The pictogram can be supplemented by text if this is considered necessary to make the sign easily understood (BS-type sign), but you must not have a safety sign that uses only text. Either type of sign can be used but different types should not be mixed. Appropriate signs should take into account the needs of those who may need to use them.

The legibility of escape signs is determined by the size of the sign, its level of illumination and the distance over which it is viewed. The use of signs within the same premises should follow a consistent design pattern or scheme. You should not rely on a few outsized signs which may encourage people to travel to a particular escape route when other more appropriate routes should be used.

In multi-occupied premises, co-operation between the respective 'responsible persons' should be sought to ensure that, as far as possible, all signs in the building conform to a single pattern or scheme.

* The Royal National Institute of the Blind estimates that only about 4% of visually impaired people are totally blind.

Figure 57: BS-type sign

Figure 58: Euro sign

Other safety signs and notices

A number of other mandatory signs such as 'Fire action' notices may also be necessary.

Fire doors that have been fitted with self-closing devices should be labelled 'Fire door – keep shut' (Figure 59) on both sides. fire-resisting doors to cupboards, stores and service ducts that are not self-closing because they are routinely kept locked should be labelled 'Fire door – keep locked' on the outside.

Figure 59: Fire door 'keep shut' notice

Signs should indicate non-automatic fire safety equipment if there is any doubt about its location, e.g. fire extinguishers that are kept in cabinets or in recesses.

A notice with the words 'Push bar to open' should be permanently displayed immediately above the push-bar on all doors fitted with a panic bolt or panic latch.

A notice with the words 'Fire escape – keep clear' should be permanently displayed at about eye level on the external face of all doors which are provided as a means of escape in case of fire and which, because they are not normally used, may become obstructed.

Staff notices

In simple premises where there are a limited number of escape routes, it may be reasonable to provide staff with verbal reminders of what they need to do if there is a fire. In some premises, you could consider providing this in a short written statement that could, for example, be delivered with staff pay slips every six months.

In multi-occupied, larger and more complex premises or where there is a high turnover of staff, a more considered approach for staff notices and instructions will be necessary. As well as positioning the fire instructions notice on escape routes adjacent to fire break-glass call points (Figure 60), put them where staff frequently assemble in the premises, e.g. the canteen and locker rooms.

Figure 60: Staff action notice

If your premises are routinely expected to accommodate people whose first language is not English you may need to consider providing instruction in more than one language. The interpretation should always convey an identical message.

Illumination

All signs and notices will need illumination to ensure they are conspicuous and legible. There are a number of options available to achieve this, such as:

- external illumination; and
- internal illumination.

The supplier or other competent person can give you further advice.

Signs or notices of the photo-luminescent type, i.e. where the active material making up the luminous parts of such signs or notices needs a period of exposure to light before they become visible in darkness (but get fainter with time), are not a substitute for appropriate emergency lighting and should only be used where other forms of illumination are present.

Further guidance

Detailed guidance on fire safety signs can be found in BS 5499-4,[30] BS 5499-5.[74] Published guidance[5,6] on compliance with health and safety legislation on signs is also available. Guidance about the use of photo-luminescent fire safety signs and notices can be found in BS 5266-6.[73]

Section 7 Further guidance on recording, planning, informing, instructing and training

7.1 Fire safety records

Keeping up-to-date records of your fire risk assessment can help you effectively manage the fire strategy for your premises and demonstrate how you are complying with fire safety law.

Even if you do not have to record the fire risk assessment, it can be helpful to keep a record of any co-operation and exchange of information made between employers and other responsible people for future reference.

In larger and more complex premises, it is best to keep a dedicated record of all maintenance of fire-protection equipment and training. There is no one 'correct' format specified for this. Suitable record books are available from trade associations and may also be available from your local enforcing authority.

In all cases the quality of records may also be regarded as a good indicator of the overall quality of the safety management structure.

Your records should be kept in a specified place on the premises (for example, in the management's office), and should include:

- details of any significant findings from the fire risk assessment and any action taken (see Part 1, Section 4.1);
- testing and checking of escape routes, including final exit locking mechanisms, such as panic devices, emergency exit devices and any electromagnetic devices;
- testing of fire-warning systems, including weekly alarm tests and periodic maintenance by a competent person;
- recording of false alarms;
- testing and maintenance of emergency lighting systems;
- testing and maintenance of fire extinguishers, hose reels and fire blankets etc.;
- if appropriate, testing and maintenance of other fire safety equipment such as fire-suppression systems and smoke control systems;
- recording and training of relevant people and fire evacuation drills;
- planning, organising, policy and implementation, monitoring, audit and review;
- maintenance and audit of any systems that are provided to help the fire and rescue service;
- the arrangements in a large multi-occupied building for a co-ordinated emergency plan or overall control of the actions you or your staff should take if there is a fire; and
- all alterations, tests, repairs and maintenance of fire safety systems, including passive systems such as fire doors.

Other issues that you may wish to record include:

- the competence, qualifications and status of the persons responsible for carrying out inspections and tests;
- the results of periodic safety audits, reviews, inspections and tests, and any remedial action taken;
- all incidents and circumstances which had the potential to cause accidents and monitor subsequent remedial actions; and
- a record of the building use, the fire prevention and protection measures in place and high-risk areas.

You should ensure that no other management decisions or policies compromise safety.

Your documentation should be available for inspection by representatives of the enforcing authority.

More detailed advice is given in BS 5588-12.[52]

Figure 61 is an example of how to record some individual stages of the process in more detail. A blank version of this form is provided in Appendix A2.

Figure 61: Example record of significant findings

Risk Assessment – Record of significant findings

Risk assessment for		Assessment undertaken by	
Company	abc stores	Date	02/02/2006
Address	2 The High St Any Town AB12 3CD	Completed by Signature	J Smith J Smith

Sheet number	Floor/area	Use
One	Ground floor sales area	Shop

Step 1 – Identify fire hazards

Sources of ignition	Sources of fuel	Sources of oxygen
Hot lamp in display area Electrical heaters Microwave/kettle in tea bay	• Fabrics and textiles • Display material • Packaging • Stationery	No additional sources

Step 2 – People at risk

6 full time staff members 2 part time staff members 60 customers at peak periods	1 Clerical assistant working in first floor office (1 wheel chair user)

Step 3 – Evaluate, remove, reduce and protect from risk

(3.1) Evaluate the risk of the fire occuring	• Hot lamps too close to display materials • 1 wall heater close to shelving/goods • Tea bay routinely use for storing packaging material
(3.2) Evaluate the risk to people from a fire starting in the premises	• Predominantly open plan so fire in sales area quickly evident • Fire in rear store could go undetected • Fire on sales floor could effect first floor office
(3.3) Remove and reduce the hazards that may cause a fire	• Replace hot lamps with overhead low wattage track lighting • Remove wall heater as other 2 heaters considered adequate • Tea bay to be moved to first floor • Packaging material to be placed in secure refuse store to rear of premises
(3.4) Remove and reduce the risks to people from a fire	The current fire precautions measures (refer to floor plan A) have been assessed in view of the findings recorded above. They are considered adequate with the following exceptions Additional measures considered necessary • Provide automatic fire detection to rear store • Fire door protecting staircase to 1st floor office requires new self closer • Ramp to rear fire exit requires repair • Fire extinguishers require maintenance • Introduce periodic staff refresher training • Replace damaged fire exit sign above rear exit

Assessment review

Assessment review date	Completed by	Signature

Review outcome (where substantial changes have occurred a new record sheet should be used)

Notes:

(1) The risk assessment record of significant findings should refer to other plans, records or other documents as necessary.

(2) The information in this record should assist you to develop an emergency plan; coordinate measures with other 'responsible persons' in the building; and to inform and train staff and inform other relevant persons.

In premises with 'engineered fire safety strategies', a fire policy manual should be provided in addition to any other records. Enforcing authorities would expect a fire engineering policy manual to conform to the structure set out in BS 7974-0 section 5: Reporting and presentation.[31]

Fire safety audit

A fire safety audit can be used alongside your fire risk assessment to identify what fire safety provisions exist in your premises.

When carrying out a review of your fire safety risk assessment, a pre-planned audit can quickly identify if there have been any significant changes which may affect the fire safety systems and highlight whether a full fire risk assessment is necessary.

Plans and specifications

Plans and specifications can be required to assist understanding of a fire risk assessment or emergency plan. Even where not needed for this purpose they can help you and your staff keep your fire risk assessment and emergency plan under review and help the fire and rescue service in the event of fire. Any symbols used should be shown on a key. Plans and specifications could include the following:

- essential structural features such as the layout of function rooms, escape doors, wall partitions, corridors, stairways etc. (including any fire-resisting structure and self-closing fire doors provided to protect the escape routes);
- location of refuges and lifts that have been designated suitable for use by disabled people and others who may need assistance to escape in case of fire;
- methods for fighting fire (details of the number, type and location of the firefighting equipment);
- location of manually-operated fire alarm call points and control equipment for fire alarms;
- location of any control rooms and any fire staff posts;
- location of any emergency lighting equipment and the exit route signs;
- location of any high-risk areas, equipment or process that must be immediately shut down by staff on hearing the fire alarm;
- location of any automatic firefighting systems, risers and sprinkler control valves;
- location of the main electrical supply switch, the main water shut-off valve and, where appropriate, the main gas or oil shut-off valves; and
- plans and specifications relating to all recent constructions.

This information should be passed on to any later users or owners of the premises.

7.2 Emergency plans

Emergency plan and contingency plans

Your emergency plan should be appropriate to your premises and could include:

- how people will be warned if there is a fire;
- what staff should do if they discover a fire;
- how the evacuation of the premises should be carried out;
- where people should assemble after they have left the premises and procedures for checking whether the premises have been evacuated;
- identification of key escape routes, how people can gain access to them and escape from them to a place of total safety;
- arrangements for fighting the fire;
- the duties and identity of staff who have specific responsibilities if there is a fire;
- arrangements for the safe evacuation of people identified as being especially at risk, such as those with disabilities, lone workers and young persons;
- any machines/appliances/processes/power supplies that need to be stopped or isolated if there is a fire;
- specific arrangements, if necessary, for high-fire-risk areas;
- contingency plans for when life safety systems such as evacuation lifts, fire-detection and warning systems, sprinklers or smoke control systems are out of order;
- how the fire and rescue service and any other necessary services will be called and who will be responsible for doing this;
- procedures for meeting the fire and rescue service on their arrival and notifying them of any special risks, e.g. the location of highly flammable materials;

- what training employees need and the arrangements for ensuring that this training is given;
- phased evacuation plans (where some areas are evacuated while others are alerted but not evacuated until later); and
- plans to deal with people once they have left the premises.

As part of your emergency plan it is good practice to prepare post-incident plans for dealing with situations that might arise such as those involving:

- unaccompanied children;
- people with personal belongings (especially valuables) still in the building;
- people wishing to rejoin friends;
- getting people away from the building (e.g. to transport); and
- inclement weather.

You should therefore prepare contingency plans to determine specific actions and/or the mobilisation of specialist resources.

Guidance on developing health and safety management policy has been published by the HSE.[32]

7.3 Information, instruction, co-operation and co-ordination

Supplying information

You must provide easily understandable information to employees, the parents of children you may employ, and to employers of other persons working in your premises about the measures in place to ensure a safe escape from the building and how they will operate, for example:

- any significant risks to staff and other relevant persons that have been identified in your fire risk assessment or any similar assessment carried out by another user and responsible person in the building;
- the fire prevention and protection measures and procedures in your premises and where they impact on staff and other relevant persons in the building;
- the procedures for fighting a fire in the premises; and
- the identity of people who have been nominated with specific responsibilities in the building.

Even if you do not have to record the fire risk assessment, it would be helpful to keep a record of any co-operation and exchange of information made between employers and other responsible people for future reference.

You need to ensure that all staff and, where necessary, other relevant persons in the building, receive appropriate information in a way that can be easily understood. This might include any special instructions to particular people who have been allocated a specific task, such as shutting down equipment or guiding people to the nearest exit.

Duties of employees to give information

Employees also have a duty to take reasonable care for their own safety and that of other people who may be affected by their activities. This includes the need for them to inform their employer of any activity that they consider would present a serious and immediate danger to their own safety and that of others.

Dangerous substances

HSE publishes guidance[8] about specific substances where appropriate information may need to be provided. If any of these, or any other substance that is not included but nevertheless presents more than a slight risk, is present in your premises, then you must provide such information to staff and others, specifically you must:

- name the substance and the risks associated with it, e.g. how to safely use or store the product to avoid creating highly flammable vapours or explosive atmospheres;
- identify any legislative provisions that may be associated with the substance;
- allow employees access to the hazardous substances safety data sheet; and
- inform the local fire and rescue service where dangerous substances are present on the premises.

> Case study
>
> A few boxes of sodium chlorate-based weedkiller, or a few litres of flammable varnish, are unlikely to need anything other than basic precautions such as a warning sign on the room or container in which they are stored. However, the storage and/or use of sizeable quantities of polystyrene pellets for the packaging of retail goods purchased by customers or significant quantities of highly flammable liquids stored in a large builders' merchant will require more comprehensive information and notification to the fire and rescue service.

Information to the fire and rescue services
In addition to providing information to the fire and rescue service when dangerous substances are present in sufficient quantities to pose an enhanced risk, it will also be helpful to inform them of any short term changes that might have an impact on their firefighting activities, e.g. in the event of temporary loss of a firefighting facility and temporary alterations.

Procedures should also include meeting and briefing the fire and rescue service when they arrive.

> Case study
>
> If the firefighting lift in a multi-storey office premises becomes defective, this should be brought to the attention of the fire and rescue service. Being unable to use this facility to tackle a fire on the upper floors might have a serious effect on the ability of firefighters to begin operations as quickly as planned. The information supplied will enable the emergency services to make adjustments to the level of the emergency response.

Instruction
You will need to carefully consider the type of instructions to staff and other people working in your premises. Written instructions must be concise, comprehensible and relevant and therefore must be reviewed and updated as new working practices and hazardous substances are introduced.

Inclusive access and employment policies mean that people with learning difficulties may now be present in a range of premises and your fire risk assessment should consider whether further instruction or guidance is necessary to ensure that your evacuation strategy is appropriate and understood by everyone.

Instructions will need to be given to people delegated to carry out particular tasks, for example:

- removing additional security, bolts, bars or chains on final exit doors before the start of business to ensure that escape routes are accessible;
- daily, weekly, quarterly and yearly checks on the range of fire safety measures (in larger premises some of the work may be contracted out to a specialist company);
- safety considerations when closing down the premises at the end of the day, e.g. removing rubbish, ensuring enough exits are available for people that remain and closing fire doors and shutters;
- leaving hazardous substances in a safe condition when evacuating the building;
- the safe storage of hazardous substances at the end of the working day; and
- ensuring everyone in large organisations with many buildings within a curtilage and a security zone know how to use internal emergency telephones.

Specific instructions may be needed about:

- how staff will help members of the public/visitors to leave the building;
- 'sweeping' of the shop floor by staff to guide people to the nearest exit when the fire alarm sounds;
- designating particular areas of a shop/office for supervisors to check that no one remains inside;
- calling the emergency services;
- carrying out evacuation roll calls;
- taking charge at the assembly area;
- meeting and directing fire engines; and
- cover arrangements when nominated people are on leave.

Co-operation and co-ordination

Where you share premises with others (this includes people who are self-employed or in partnership), each responsible person, i.e. each employer, owner or other person who has control over any part of the premises, will need to co-operate and co-ordinate the findings of their separate fire risk assessments to ensure the fire precautions and protection measures are effective throughout the building. This could include:

- co-ordinating an emergency plan (see Step 4.2 for features of an emergency plan);
- identifying the nature of any risks and how they may affect others in or about the premises;
- identifying any fire-prevention and protection measures;
- identifying any measures to mitigate the effects of a fire; and
- arranging any contacts with external emergency services and calling the fire and rescue service.

> Case study
>
> Company A stores and sells a wide range of timber products including fence panels, garden sheds and timber decking and operates as a franchise within a garden centre owned and occupied by Company B. Company A's premises are in the open air where significant stockpiles of timber fencing are stored on 8 m-high steel racking on the boundary of the garden centre. Should a fire start there is a significant risk that it will spread to other parts of the garden centre and to other adjoining premises, endangering people in both premises and in the immediate vicinity. The responsible person (Company A) should ensure that there is enough space between stacks of timber products and the boundary to ensure as far as possible that any fire is confined to the stack and would not spread and endanger other people.
> To do this, companies A and B will need to co-operate and co-ordinate as necessary.

7.4 Fire safety training

Staff training

The actions of staff if there is a fire are likely to be crucial to their safety and that of other people in the premises. All staff should receive basic fire safety induction training and attend refresher sessions at pre-determined intervals.

You should ensure that all staff and contractors are told about the emergency plan and are shown the escape routes.

The training should take account of the findings of the fire risk assessment and be easily understood by all those attending. It should include the role that those members of staff will be expected to carry out if a fire occurs. This may vary in large premises, with some staff being appointed as fire marshals or being given some other particular role for which additional training will be required.

In addition to the guidance given in Part 1, Step 4.4, as a minimum all staff should receive training about:

- the items listed in your emergency plan;
- the importance of fire doors and other basic fire-prevention measures;
- where relevant, the appropriate use of firefighting equipment;
- the importance of reporting to the assembly area;
- exit routes and the operation of exit devices, including physically walking these routes;
- general matters such as permitted smoking areas or restrictions on cooking other than in designated areas; and
- assisting disabled persons where necessary.

Training is necessary:

- when staff start employment or are transferred into the premises;
- when changes have been made to the emergency plan and the preventive and protective measures;
- where working practices and processes or people's responsibilities change;
- to take account of any changed risks to the safety of staff or other relevant persons;

- to ensure that staff know what they have to do to safeguard themselves and others on the premises;
- where staff are expected to assist disabled persons; and
- if a member of staff may take on the role of duty manager.

Training should be repeated as often as necessary and should take place during working hours.

Whatever training you decide is necessary to support your fire safety strategy and emergency plan, it should be verifiable.

Enforcing authorities may want to examine records as evidence that adequate training has been given.

Fire marshals

Staff expected to undertake the role of fire marshals (often called fire wardens) would require more comprehensive training. Their role may include:

- helping those on the premises to leave;
- checking the premises to ensure everyone has left;
- using firefighting equipment if safe to do so;
- liaising with the fire and rescue service on arrival;
- shutting down vital or dangerous equipment; and
- performing a supervisory/managing role in any fire situation.

Training for this role may include:

- detailed knowledge of the fire safety strategy of the premises;
- awareness of human behaviour in fires;
- how to encourage others to use the most appropriate escape route;
- how to search safely and recognise areas that are unsafe to enter;
- the difficulties that some people, particularly if disabled, may have in escaping and any special evacuation arrangements that have been pre-planned;
- additional training in the use of firefighting equipment;
- an understanding of the purpose of any fixed firefighting equipment such as sprinklers or gas flooding systems; and
- reporting of faults, incidents and near misses.

Fire drills

Once the emergency plan has been developed and training given, you will need to evaluate its effectiveness. The best way to do this is to perform a fire drill. This should be carried out at least annually or as determined by your fire risk assessment. If you have a high staff turnover, you may need to carry them out more often.

A well-planned and executed fire drill will confirm understanding of the training and provide helpful information for future training. The responsible person should determine the possible objectives of the drill such as to:

- identify any weaknesses in the evacuation strategy;
- test the procedure following any recent alteration or changes to working practices;
- familiarise new members of staff with procedures; and
- test the arrangements for disabled people.

Who should take part?

Within each building the evacuation should be for all occupants except those who may need to ensure the security of the premises, or people who, on a risk-assessed basis, are required to remain with particular equipment or processes that cannot be closed down.

Premises that consist of several buildings on the same site should be dealt with one building at a time over an appropriate period unless the emergency procedure dictates otherwise.

Where appropriate, you may find it helpful to include members of the public in your fire drill – ensuring that all necessary health and safety issues are addressed before you do so.

Carrying out the drill

For premises that have more than one escape route, the escape plan should be designed to evacuate all people on the assumption that one exit or stairway is unavailable because of the fire. This could be simulated by a designated person being located at a suitable point on an exit route. Applying this scenario to different escape routes at each fire drill will encourage individuals to use alternative escape routes which they may not normally use.

When carrying out the drill you might find it helpful to:

- circulate details concerning the drill and inform all staff of their duty to participate. It may not be beneficial to have 'surprise drills' as the health and safety risks introduced may outweigh the benefits;
- ensure that equipment can be safely left;
- nominate observers;
- inform the alarm receiving centre if the fire-warning system is monitored (if the fire and rescue service is normally called directly from your premises, ensure that this does not happen);
- inform visitors and members of the public if they are present; and
- ask a member of staff at random to set off the alarm by operating the nearest alarm call point using the test key. This will indicate the level of knowledge regarding the location of the nearest call point.

More detailed information on fire drills and test evacuations are given in BS 5588-12.[52]

The roll call/checking the premises have been evacuated

Where possible, you should ensure that a roll call is carried out as soon as possible at the designated assembly point(s), and/or receive reports from wardens designated to 'sweep' the premises. You should note any people who are unaccounted for. In a real evacuation this information will need to be passed to the fire and rescue service on arrival.

Check that people have assembled at the evacuation point.

Once the roll call is complete or all reports have been received, allow people to return to the building. If the fire-warning system is monitored inform the alarm receiving centre that the drill has now been completed and record the outcomes of the drill.

Monitoring and debrief

Throughout the drill the responsible person and nominated observers should pay particular attention to:

- communication difficulties with regard to the roll call and establishing that everyone is accounted for;
- the use of the nearest available escape routes as opposed to common circulation routes;
- difficulties with the opening of final exit doors;
- difficulties experienced by people with disabilities;
- the roles of specified people, e.g. fire wardens;
- inappropriate actions, e.g. stopping to collect personal items, attempting to use lifts etc.; and
- windows and doors not being closed as people leave.

On-the-spot debriefs are useful to discuss the fire drill, encouraging feedback from everybody. Later, reports from fire wardens and observations from people should be collated and reviewed. Any conclusions and remedial actions should be recorded and implemented.

Section 8 Quality assurance of fire protection equipment and installation

Fire protection products and related services should be fit for their purpose and properly installed and maintained in accordance with the manufacturer's instructions or the relevant British Standard.

Third-party certification schemes for fire protection products and related services are an effective means of providing the fullest possible assurances, offering a level of quality, reliability and safety that non-certificated products may lack. This does not mean goods and services that are not third-party approved are less reliable, but there is no obvious way in which this can be demonstrated.

Third-party quality assurance can offer comfort both as a means of satisfying you that goods and services you have purchased are fit for purpose, and as a means of demonstrating that you have complied with the law.

However, to ensure the level of assurance offered by third party schemes, you should always check whether the company you employ sub-contracts work to others. If they do, you will want to check that the sub-contractors are subject to the same level of checks of quality and competence as the company you are employing.

Your local fire and rescue service, fire trade associations or your own trade association may be able to provide further details about third-party quality assurance schemes and the various organisations that administer them.

Appendix A

A.1 Example fire safety maintenance checklist

A fire safety maintenance checklist can be used as a means of supporting your fire safety policy. This example list is not intended to be comprehensive and should not be used as a substitute for carrying out a fire risk assessment.

You can moidfy the example where necessary, to fit your premises and may need to incorporate the recommendations of manufacturers and installers of the fire safety equipment/systems that you may have installed in your premises.

Any ticks in the grey boxes should result in further investigation and appropriate action as necessary. In larger and more complex premises you may need to seek the assistance of a competent person to carry out some of the checks.

	Yes	No	N/A	Comments
Daily checks (not normally recorded)				
Escape routes				
Can all fire exits be opened immediately and easily?	☐	☐	☐	
Are fire doors clear of obstructions?	☐	☐	☐	
Are escape routes clear?	☐	☐	☐	
Fire warning systems				
Is the indicator panel showing 'normal'?	☐	☐	☐	
Are whistles, gongs or air horns in place?	☐	☐	☐	
Escape lighting				
Are luminaires and exit signs in good condition and undamaged?	☐	☐	☐	
Is emergency lighting and sign lighting working correctly?	☐	☐	☐	
Firefighting equipment				
Are all fire extinguishers in place?	☐	☐	☐	
Are fire extinguishers clearly visible?	☐	☐	☐	
Are vehicles blocking fire hydrants or access to them?	☐	☐	☐	
Weekly checks				
Escape routes				
Do all emergency fastening devices to fire exits (push bars and pads, etc.) work correctly?	☐	☐	☐	
Are external routes clear and safe?	☐	☐	☐	
Fire warning systems				
Does testing a manual call point send a signal to the indicator panel? (Disconnect the link to the receiving centre or tell them you are doing a test.)	☐	☐	☐	
Did the alarm system work correctly when tested?	☐	☐	☐	
Did staff and other people hear the fire alarm?	☐	☐	☐	
Did any linked fire protection systems operate correctly? (e.g. magnetic door holder released, smoke curtains drop)	☐	☐	☐	

	Yes	No	N/A	Comments
Weekly checks *continued*				
Do all visual alarms and/or vibrating alarms and pagers (as applicable) work?	☐	☐	☐	
Do voice alarm systems work correctly? Was the message understood?	☐	☐	☐	
Escape lighting				
Are charging indicators (if fitted) visible?	☐	☐	☐	
Firefighting equipment				
Is all equipment in good condition?	☐	☐	☐	
Additional items from manufacturer's recommendations.	☐	☐	☐	
Monthly checks				
Escape routes				
Do all electronic release mechanisms on escape doors work correctly? Do they 'fail safe' in the open position?	☐	☐	☐	
Do all automatic opening doors on escape routes 'fail safe' in the open position?	☐	☐	☐	
Are fire door seals and self-closing devices in good condition?	☐	☐	☐	
Do all roller shutters provided for fire compartmentation work correctly?	☐	☐	☐	
Are external escape stairs safe?	☐	☐	☐	
Do all internal self-closing fire doors work correctly?	☐	☐	☐	
Escape lighting				
Do all luminaires and exit signs function correctly when tested?	☐	☐	☐	
Have all emergency generators been tested? (Normally run for one hour.)	☐	☐	☐	
Firefighting equipment				
Is the pressure in 'stored pressure' fire extinguishers correct?	☐	☐	☐	
Additional items from manufacturer's recommendations.	☐	☐	☐	
Three-monthly checks				
General				
Are any emergency water tanks/ponds at their normal capacity?	☐	☐	☐	
Are vehicles blocking fire hydrants or access to them?	☐	☐	☐	
Additional items from manufacturer's recommendations.	☐	☐	☐	
Six-monthly checks				
General				
Has any firefighting or emergency evacuation lift been tested by a competent person?	☐	☐	☐	
Has any sprinkler system been tested by a competent person?	☐	☐	☐	
Have the release and closing mechanisms of any fire-resisting compartment doors and shutters been tested by a competent person?	☐	☐	☐	
Fire warning system				
Has the system been checked by a competent person?	☐	☐	☐	

	Yes	No	N/A	Comments
Six-monthly checks *continued*				
Escape lighting				
Do all luminaires operate on test for one third of their rated value?	☐	☐	☐	
Additional items from manufacturer's recommendations.	☐	☐	☐	
Annual checks				
Escape routes				
Do all self-closing fire doors fit correctly?	☐	☐	☐	
Is escape route compartmentation in good repair?	☐	☐	☐	
Escape lighting				
Do all luminaires operate on test for their full rated duration?	☐	☐	☐	
Has the system been checked by a competent person?	☐	☐	☐	
Firefighting equipment				
Has all firefighting equipment been checked by a competent person?	☐	☐	☐	
Miscellaneous				
Has any dry/wet rising fire main been tested by a competent person?	☐	☐	☐	
Has the smoke and heat ventilation system been tested by a competent person?	☐	☐	☐	
Has external access for the fire service been checked for ongoing availability?	☐	☐	☐	
Have any firefighters' switches been tested?	☐	☐	☐	
Has the fire hydrant bypass flow valve control been tested by a competent person?	☐	☐	☐	
Are any necessary fire engine direction signs in place?	☐	☐	☐	

A2 Example form for recording significant findings

Risk Assessment – Record of significant findings		
Risk assessment for		**Assessment undertaken by**
Company		Date
Address		Completed by
		Signature
Sheet number	**Floor/area**	**Use**
Step 1 – Identify fire hazards		
Sources of ignition	**Sources of fuel**	**Sources of oxygen**
Step 2 – People at risk		
Step 3 – Evaluate, remove, reduce and protect from risk		
(3.1) Evaluate the risk of the fire occuring		
(3.2) Evaluate the risk to people from a fire starting in the premises		
(3.3) Remove and reduce the hazards that may cause a fire		
(3.4) Remove and reduce the risks to people from a fire		
Assessment review		
Assessment review date	**Completed by**	**Signature**
Review outcome (where substantial changes have occurred a new record sheet should be used)		

Notes:
(1) The risk assessment record of significant findings should refer to other plans, records or other documents as necessary.
(2) The information in this record should assist you to develop an emergency plan; coordinate measures with other 'responsible persons' in the building; and to inform and train staff and inform other relevant persons.

Appendix B

Technical information on fire-resisting separation, fire doors and door fastenings

B1 Fire-resisting separation

General

The materials from which your premises are constructed may determine the speed with which a fire may spread, affecting the escape routes that people will use. A fire starting in a building constructed mainly from readily combustible material will spread faster than one where modern fire-resisting construction materials have been used. Where non-combustible materials are used and the internal partitions are made from fire-resisting materials, the fire will be contained for a longer period, allowing more time for the occupants to escape.

Because of the requirements of the Building Regulations you will probably already have some walls and floors that are fire-resisting and limitations on the surface finishes to certain walls and ceilings.

You will need to consider whether the standard of fire resistance and surface finishing in the escape routes is satisfactory, has been affected by wear and tear or alterations and whether any improvements are necessary.

The following paragraphs give basic information on how fire-resisting construction can provide up to 30 minutes protection to escape routes. This is the standard recommended for most situations. If you are still unsure of the level of fire resistance which is necessary after reading this information, you should consult a fire safety expert.

Fire-resisting construction

The fire resistance of a wall or floor is dependent on the quality of construction and materials used. Common examples of types of construction that provide 30-minute fire resistance to escape routes if constructed to the above standards are:

- internal framed construction wall, non-load bearing, consisting of 72mm x 37mm timber studs at 600mm centres and faced with 12.5mm of plasterboard with all joints taped and filled (see Figure 62);
- internal framed construction, non-load bearing, consisting of channel section steel studs at 600mm centres faced with 12.5mm of plasterboard with all joints taped and filled; and
- masonry cavity wall consisting of solid bricks of clay, brick earth, shale, concrete or calcium silicate, with a minimum thickness of 90mm on each leaf.

There are other methods and products available which will achieve the required standard of fire resistance and may be more appropriate for the existing construction in your premises. If there is any doubt about how your building is constructed, then ask for further advice from a competent person.

Figure 62: Fire-resisting construction

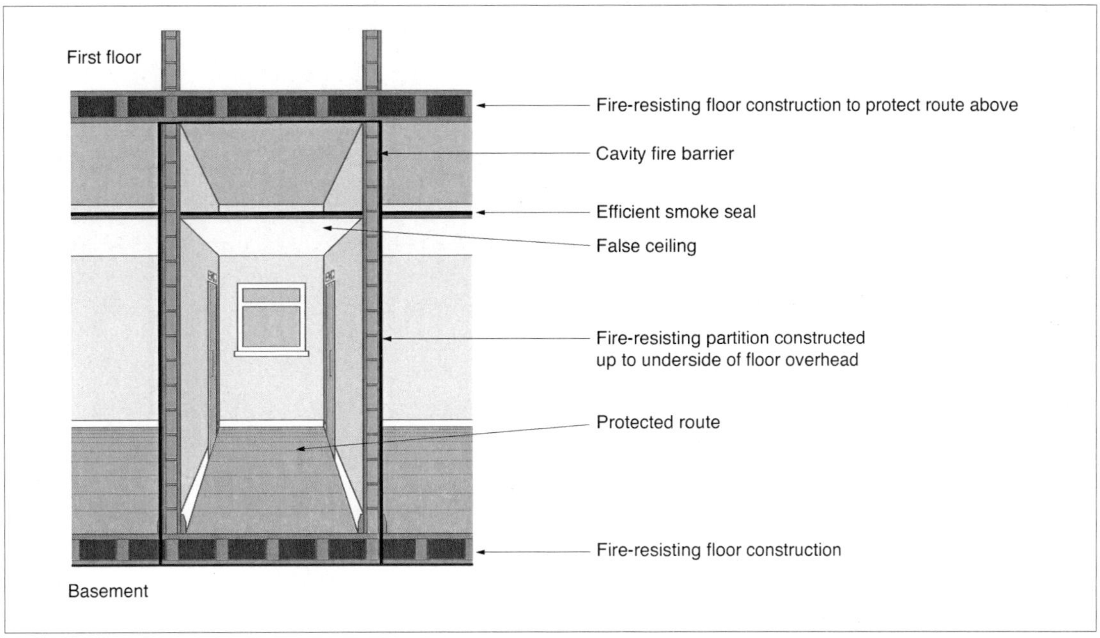

Fire-resisting floors

The fire resistance of floors will depend on the existing floor construction as well as the type of ceiling finish beneath. If you need to upgrade the fire resistance of your floor it may not be desirable to apply additional fire resistance to the underside of an existing ornate ceiling. In older buildings there may be a requirement to provide fire resistance between beams and joists.

A typical example of a 30-minute fire-resisting timber floor is tongue and groove softwood of not less than 15mm finished thickness on 37mm timber joists, with a ceiling below of one layer of plasterboard to a thickness of 12.5mm with joints taped and filled and backed by supporting timber.

There are other, equally valid, methods and products available for upgrading floors. If you are in any doubt you should ask the advice of a competent person and ensure that the product is installed in accordance with instructions from the manufacturer or supplier.

Fire-resisting glazing

The most common type of fire-resisting glazing is 6mm Georgian wired glazing, which is easily identifiable. Clear fire-resisting glazing is available and can quickly be identified by a mark etched into the glass, usually in the corner of the glazed panel, to confirm its fire-resisting standard. Although this is not compulsory, the marking of glass is supported by the Glass and Glazing Federation; you should check whether the glazing will be marked accordingly before purchase. The glazing should have been installed in accordance with the manufacturer's instructions and to the appropriate standard,[66] to ensure that its fire-resisting properties are maintained.

The performance of glazed systems in terms of fire resistance and external fire exposure should, wherever possible, be confirmed by test evidence. Alternatively, where there is a lack of test information, ask for an assessment of the proposed construction from suitably qualified people.

Fire separation of voids

A common problem encountered with fire separation is fire-resisting partitions which do not extend above false ceilings to true ceiling height. This may result in unseen fire spread and a loss of vital protection to the escape routes. It is important therefore to carefully check all such partitions have been installed correctly.

CLASP and SCOLA type construction

CLASP (Consortium of Local Authorities Special Programme) and SCOLA (Second Consortium of Local Authorities) are total or systematic methods of construction that were developed to provide consistent building quality, while reducing the need for traditional skilled labour.

They consist of a metal frame upon which structural panels are fixed. This results in hidden voids through which fire may spread. It is important that cavity barriers that restrict the spread of fire are installed appropriately, especially to walls and floors that need to be fire-resisting. If you are in any doubt as to whether any remedial work will be required, then ask for advice from a competent person.

Breaching fire separation

To ensure effective protection against fire, walls and floors providing fire separation must form a complete barrier, with an equivalent level of fire resistance provided to any openings such as doors, ventilation ducts, pipe passages or refuse chutes.

The passing of services such as heating pipes or electrical cables through fire-resisting partitions leaves gaps through which fire and smoke may spread. This should be rectified by suitable fire stopping and there are many proprietary products available to suit particular types of construction. Such products should be installed by competent contractors.

Décor and surface finishes of walls, ceilings and escape routes

The materials used to line walls and ceilings can contribute significantly to the spread of flame across their surface. Most materials that are used as surface linings will fall into one of three classes of surface spread of flame. The following are common examples of acceptable materials for various situations:

Class 0: Materials suitable for circulation spaces and escape routes

- Such materials include brickwork, blockwork, concrete, ceramic tiles, plaster finishes (including rendering on wood or metal lathes), wood-wool cement slabs and mineral fibre tiles or sheets with cement or resin binding.

Note: Additional finishes to these surfaces may be detrimental to the fire performance of the surface and if there is any doubt about this then consult the manufacturer of the finish.

Class 1: Materials suitable for use in all rooms but not on escape routes

- Such materials include all the Class 0 materials referred to above. Additionally, timber, hardboard, blockboard, particle board, heavy flock wallpapers and thermosetting plastics will be suitable if flame-retardant treated to achieve a Class 1 standard.

Class 3: Materials suitable for use in rooms of less than 30m^2

- Such materials include all those referred to in Class 1, including those that have not been flame-retardant treated and certain dense timber or plywood and standard glass-reinforced polyesters.

The equivalent European classification standard will also be acceptable. Further details about internal linings and classifications are available in Approved Document B.[24] Appropriate testing procedures are detailed in BS 476-7[33] and where appropriate BS EN 13501-1.[34]

Further guidance on types of fire-resisting construction has been published by the Building Research Establishment.[35]

B2 Fire-resisting doors

Requirements of a fire-resisting door

Effective fire-resisting doors (see Figure 63) are vital to ensure that the occupants can evacuate to a place of safety. Correctly specified and well-fitted doors will hold back fire and smoke, preventing escape routes becoming unusable, as well as preventing the fire spreading from one area to another.

Fire-resisting doors are necessary in any doorway located in a fire-resisting structure. Most internal doors are constructed of timber. These will give some limited protection against fire spread, but only a purpose-built fire-resisting door that has been tested to an approved standard will provide the necessary protection. Metal fire-resisting doors are also available and specific guidance for these follows.

All fire-resisting doors are rated by their performance when tested to an appropriate standard. The level of protection provided by the door is measured, primarily by determining the time taken for a fire to breach the integrity (E), of the door assembly, together with its resistance to the passage of hot gases and flame.

It may be possible to upgrade the fire resistance of existing doors. Further information is available from the Building Research Establishment[67] or Timber Research and Development Association.[68]

Figure 63: A fire-resisting and smoke-stopping door

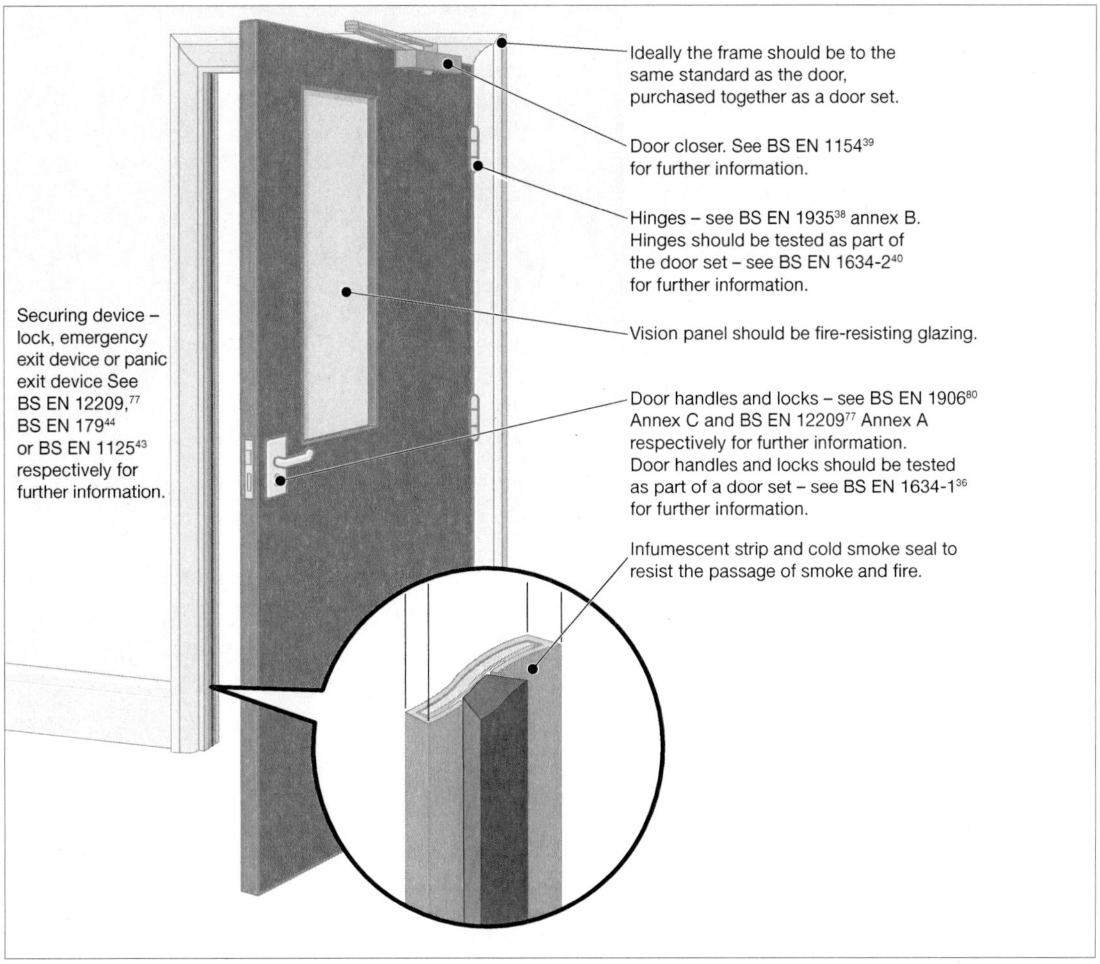

Timber fire-resisting doors require a gap of 2-4mm between the door leaf and the frame. However larger gaps may be necessary to ensure that the door closes flush into its frame when smoke seals are fitted. Further information is available in BS 4787-1.[69] For fire-resisting purposes the gap is normally protected by installing an intumescent seal, in either the door or, preferably, the frame. The intumescent seal expands in the early stages of a fire and enhances the protection given by the door. Additional smoke seals will restrict the spread of smoke at ambient temperatures. Doors fitted with smoke seals, either incorporated in the intumescent seal or fitted separately, have their classification code suffixed with a 'S'.

The principal fire-resisting door categories are:

- E20 fire-resisting door providing 20 minutes fire resistance (or equivalent FD 20S). (Note: Many suppliers no longer provide an E20 type fire-resisting door.)
- E30 fire-resisting door providing 30 minutes fire resistance (or equivalent FD 30S).
- E60 fire-resisting door providing 60 minutes fire resistance (or equivalent FD 60S).

Timber fire-resisting doors are available that will provide up to 120 minutes fire resistance but their use is limited to more specialised conditions that are beyond the scope of this guidance.

Metal fire-resisting doors

Although the majority of fire-resisting doors are made from timber, metal fire-resisting doors, which meet the appropriate standard, can often be used for the same purpose. However, there are situations where they are more appropriate. The majority of metal fire-resisting door manufacturers will require the use of bespoke frames and hardware for their door sets.

See BS EN 1634-1[36] and BS 476-22[37] for more information.

For detailed guidance refer to Approved Document B.[24]

Glazing in fire-resisting doors

Although glazing provides additional safety in everyday use and can enhance the appearance of fire-resisting doors, it should never reduce the fire resistance of the door. The opening provided in the door for the fire-resisting glazing unit(s) and the fitting of the beading are critical, and should only be entrusted to a competent person. In nearly all cases the door and glazing should be purchased from a reputable supplier who can provide documentary evidence that the door continues to achieve the required rating.

Fire-resisting door furniture

Hinges

To ensure compliance with their rated fire performance, fire-resisting doors need to be hung with the correct number, size and quality of hinges. Normally a minimum of three hinges are needed, however the manufacturer's instructions should be closely followed. BS EN 1935[38] including Annex B, is the appropriate standard.

Alternative door mountings

Although the most common method of hanging a door is to use single axis hinges, alternative methods are employed where the door is required to be double swing or mounted on pivots for other reasons.

Floor mounted controlled door closing devices are the most common method regularly found with timber, glass and steel doors while transom mounted devices are commonly used with aluminium sections. In each case reference should be made to the fire test report for details as to compliance with the composition of the door assembly including the door mounting conditions.

Self-closing devices

All fire-resisting doors, other than those to locked cupboards and service ducts, should be fitted with an appropriately controlled self-closing device that will effectively close the door from any angle. In certain circumstances, concealed, jamb-mounted closing devices may be specified and in these cases should be capable of closing the door from any angle and against any latch fitted to the door; spring hinges are unlikely to be suitable. Further information is available in BS EN 1154.[39]

Rising butt hinges are not suitable for use as a self-closing device due to their inability to close and latch the door from any angle.

Automatic door hold-open/release devices for self-closing fire doors

These devices are designed to hold open self-closing fire doors or allow them to swing free during normal use. In the event of a fire alarm the device will then release the door automatically, allowing the self-closing mechanism to close the door.

Such devices are particularly useful in situations where self-closing doors on escape routes are used regularly by significant numbers of people, or by people with impaired mobility who may have difficulty in opening the doors.

Typical examples of such devices include:

- electro-magnetic devices fitted to the fire-resisting door which release when the fire detection and warning system operates, allowing a separate self-closer to close the door;
- electro-magnetic devices within the controlled door closing device which function on the operation of the fire detection and warning system; and
- 'free swing' controlled door closing devices, which operate by allowing the door leaf to work independently of the closing device in normal conditions. An electro-magnetic device within the spring mechanism linked to the fire detection and warning system ensures that the door closes on the operation of the system.

Note: Free swing devices may not be suitable in some situations, such as corridors, where draughts are a problem and the doors are likely to swing uncontrolled, causing possible difficulty or injury to certain people e.g. those with certain disabilities, the elderly and frail, or young children.

Automatic door hold open/release devices fitted to doors protecting escape routes should only be installed in conjunction with an automatic fire detection and warning system incorporating smoke detectors, that is designed to protect the escape routes in the building (see Part 2, Section 2).

In all cases the automatic device should release the fire-resisting door allowing it to close effectively within its frame when any of the following conditions occur:

- the detection of smoke by an automatic detector;

- the actuation of the fire detection and alarm system by manual means e.g. operation of break glass call point;
- any failure of the fire detection and alarm system; or
- any electrical power failure.

Other devices, including self-contained devices which perform a similar function, that are not connected directly to a fire alarm system and are not therefore able to meet the above criteria are available and may be acceptable where a site specific risk assessment can show that they are appropriate. Such devices are unlikely to be suitable for use on doors protecting single stairways or other critical means of escape.

In all cases where a door hold open device is used it should be possible to close the door manually.

A site specific risk assessment should be undertaken before any type of automatic door hold open/release device is installed. If you are unsure about the suitability of such devices in your premises, you should seek the advice of a competent person.

Further guidance about automatic door hold open/release devices is given in BS EN 1155[70] or BS 5839-3.[41]

Door co-ordinators

Where pairs of doors with rebated meeting stiles are installed it is critical that the correct closing order is maintained. Door coordinators to BS EN 1158[71] should be fitted and fully operational in all cases where the doors are self-closing.

Installation and workmanship

The reliability and performance of correctly specified fire-resisting doors can be undermined by inadequate installation. It is important that installers with the necessary level of skill and knowledge are used. Accreditation schemes for installers of fire-resisting doors are available.

Fire-resisting doors and shutters will require routine maintenance, particularly to power operation and release and closing mechanisms.

Further information is available on fire-resisting doors in BS 8214.[42] If you are unsure about the quality, the effectiveness or the fitting of your fire-resisting doors consult a fire safety expert.

For further guidance on the selection and maintenance of door furniture suitable for use on timber fire resisting and escape doors refer to The Building Hardware Industry Federation (BHIF) Code of Practice – Hardware for Timber Fire and Escape Doors.[72]

B3 Door-fastening devices

The relationship between the securing of doors against unwanted entry and the ability to escape through them easily in an emergency has often proved problematical. Careful planning and the use of quality materials remain the most effective means of satisfying both of these objectives.

Any device that impedes people making good their escape, either by being unnecessarily complicated to manipulate or not being readily openable, will not be acceptable.

Guidance on fire exits starts from the position that doors on escape routes should not be fitted with any locking devices (electrically operated or otherwise). However, it is accepted that in many cases the need for security will require some form of device that prevents unlimited access, but still enables the occupants of a building or area to open the door easily if there is a fire. These devices can take many forms but, in the majority of cases, premises where there are members of the public present or others who are not familiar with the building should use panic exit bar devices (i.e. push bars or touch bars). See BS EN 1125[43] for further information.

Premises that have limited numbers of staff or others who are familiar with the building and where panic is not likely may use alternative devices (i.e. push pads or lever handles). See BS EN 179[44] for further information.

In some larger premises, when only staff are on the premises and there is a security issue, it may be acceptable to restrict the number of emergency exits immediately available, e.g. when only security staff are present at night, or prior to opening the premises in the morning. Staff should be made fully aware of any restrictions and the number of exits not immediately available should be limited.

Electrical locking devices

Electrically operated entry control devices have been developed for use as locking devices on fire exits. They fall into two main categories, electromechanical and electromagnetic.

- Electromechanical devices

  Electromechanical devices comprise electromechanical lock keeps and draw bolts, which can be controlled by people inside the premises by entering a code or by using 'smart cards', which have been adapted to control the exit from certain areas. These devices have been fitted in many premises and may be linked to the fire-detection and/or warning system. Experience has shown that these devices can fail to open in a number of ways. They are dependent on a spring mechanism to return the lock keep or draw bolt(s) and are liable to jam when pressure is applied to the door. It is also relatively easy to fit them incorrectly. Electromechanical locking devices are normally unacceptable on escape doors, unless they are fitted with a manual means of overriding the locking mechanism, such as a push bar, push pad or lever handle or they do not rely on a spring mechanism, fail-safe open and are not affected by pressure, in which case the criteria for electromagnetic devices should be applied.

- Electromagnetic devices

  These devices comprise a magnet and a simple fixed retaining plate with no moving parts and are therefore generally considered to be more reliable due to their inherent 'fail-safe unlocked' operation. Electromagnetic locking devices go some way to addressing the particular concerns surrounding electromechanical locking systems. The release of this type of device is controlled by the interruption of electrical current to an electromagnet either manually via a switch or other means, break-glass point (typically coloured green), or by linking to the fire-warning and/or detection system of the premises.

Time-delay devices on escape routes

A further development is the fitting of a time-delay system to the electronic door-locking device. This delays the actual opening of an exit door for a variable period following operation of the panic bar or other exit device. Periods of between five and 60 seconds can be pre-set at the manufacturing stage or can be adjusted when fitted. These are not usually acceptable for use by members of the public. However, they may be acceptable for use by staff who are familiar with their operation and are suitably trained in their use.

Management of electronic door-control devices including time delays

The use of such devices may be accepted by enforcing authorities if the responsible person can demonstrate, through a suitable risk assessment for each individual door, both the need and the adequate management controls to ensure that people can escape safely from the premises. In particular:

- Access control should not be confused with exit control. Many devices are available which control the access to the premises but retain the immediate escape facility from the premises.
- In public areas, when push bars are operated on escape doors, they should release the electromagnetic locks immediately and allow the exit doors to open.
- The requirement for exit control should be carefully assessed and should not be seen as a substitute for good management of the employees and occupants.
- All other alternatives should have been explored/evaluated prior to using these devices to ensure they do not affect the safety of occupants.
- The device should be connected to the fire warning and/or detection system.
- The device should incorporate a bypass circuit for immediate release on activation of the fire warning and/or detection system.
- Each door should be fitted with a single securing device.
- The emergency exit doors should be clearly labelled about how to operate them.
- Adequate control measures should be put in place to ensure the safety of the occupants.

The use of electronic door-locking devices should be considered with particular care in premises with a number of different occupancies. The management of a complicated system of evacuation for many different groups is unlikely to be practicable.

The technical standards in respect of sourcing, maintaining and testing must be extremely high.

When part of the management control system involves trained personnel helping others at these doors, it is vital to ensure these people are available.

The use of exit control devices should not be considered where the number of trained staff is low or where members of the public would be expected to operate the devices without help.

In premises where there may be large numbers of people, the devices should only be considered when linked to a comprehensive automatic fire-detection and warning system in accordance with BS 5839-1.[16] There should be an additional means of manually overriding the locking device at each such exit (typically a green break-glass point).

The use of time-delay systems that prevent the opening of emergency exits for a pre-set time are primarily used to improve security. These add a further layer of complexity to the fire strategy and should not be considered in public areas. They should only be used in non-public areas when all other options such as relocating valuable stock or exterior boundary management have been addressed. Their value in preventing theft is likely to be transient as the use of the manual override becomes more widely known.

British Standard 8220[45] gives further advice on security in buildings and while this standard does refer to electronic locking devices, it also acknowledges that the balance must remain on the side of emergency escape rather than security.

Appendix C
Historic buildings

General considerations

This appendix offers additional information about listed and historical buildings.

Fire risk assessments conducted for a office or shop which is within a listed or historic building will need to endeavour to strike a balance between ensuring sufficient fire safety measures are in place for the safety of people, yet avoid extensive alterations and helping to maintain the character of the building.

As well as the fire risk assessment it is recommended that a general fire policy statement and manual is compiled. A person must be nominated to take responsibility for all aspects of fire safety. Usually the person charged with the management and control of the premises will be the 'responsible person' under the Order.[1]

The advice and/or consent of a building control body or any other relevant bodies (e.g. English Heritage) should form part of any fire risk assessment that impacts on the character of the building (e.g. replacement of doors, fittings, wooden panelling and decor) or material changes to existing escape routes. An ideal solution is one that is reversible, enabling the historic elements to be reinstated.

A fire safety adviser will be able to suggest alternatives to conventional fire precautions, such as:

- a fire engineering solution;
- upgrading existing doors and partitions in a sympathetic manner to improve their fire resistance; and
- considering the installation of specialist fire-detection or suppression systems.

Should the design and nature of the historic building preclude the introduction of conventional fire safety features, it will be necessary to manage the building in such a way that:

- limits the number of occupants, either staff or members of the public, inside the building;
- limits activities in the building; and
- provides adequate supervision within the building.

Liaison with the fire and rescue service

The responsible person will need to ensure effective liaison with the fire and rescue service to enable them to carry out firefighting operations. These may include information on:

- the provision of water supplies, seasonal ponds, lakes and underground tanks, and any associated pumps;
- difficult access for fire engines;
- particular hazards in the construction features of the building (including asbestos);
- the use of combustible under floor insulation;
- underground vaults ducts and voids where fire may spread unchecked;
- worn stone slabs in stairway construction; and
- the presence of cast iron columns and wrought iron beams.

Emergency planning

An important consideration for the owners and trustees is the protection of valuable artefacts and paintings from the effects of fire. However, the efficient evacuation of all occupants must take precedence over procedures for limiting damage to property and contents. Salvage work should be limited to those parts of the building not directly affected by the fire.

Fire wardens and others tasked with carrying out salvage work should have received formal training, adequate protection and be fully briefed about the health and safety risk assessment carried out to identify the dangers associated with this activity. Further detailed advice on fire safety in historic buildings can be found in the following publications:

- BS 7913: Guide to the principles of the conservation of historic buildings, British Standards Institution.
- Heritage under fire: A guide to the protection of historic buildings, Fire Protection Association (for the UK Working Party on Fire Safety in Historic Buildings) 1991, ISBN 0902167944.
- The Installation of Sprinkler Systems in Historic Buildings (Historic Scotland Technical Advice Note S.), Fire Protection Association (TCRE Division/Scottish Conservation Bureau, Hist.) 1998, ISBN 1 900168 63 4.
- Fire Protection Measures in Scottish Historic Buildings: Advice on Measures Required to Minimise the Likelihood of Fire Starting and to Alleviate the Destructive Consequences of Fire in Historic Buildings (Technical Advice Note), TCRE Division/Scottish Conservation Bureau, Hist. 1997, ISBN 1 900168 41 3.
- Fire Risk Management in Heritage Buildings (Technical Advice Note), TCRE Division/Scottish Conservation Bureau, Hist. 2001, ISBN 1 900168 71 5.
- Summary and conclusions of the report into fire protection measures for the Royal Palaces by Sir Alan Bailey. Following the Windsor Castle fire, 1992.
- The fire at Upton Park. Published by the National Trust.
- Timber panelled doors and fire. Published by English Heritage.
- Fire safety in historic town centres. Published by English Heritage and Cheshire Fire and Rescue Service.

Appendix D

Glossary

These definitions are provided to assist the responsible person in understanding some of the technical terms used in this guide. They are not exhaustive and more precise definitions may be available in other guidance.

Term	Definition
Access room	A room through which the only escape route from an inner room passes.
Accommodation stairway	A stairway, additional to that required for means of escape purposes, provided for the convenience of occupants.
Alterations notice	If your premises are considered by the enforcing authority to be high risk, they may issue an alterations notice that requires you to inform them before making any material alterations to your premises.
Alternative escape route	Escape routes sufficiently separated by either direction and space, or by fire-resisting construction to ensure that one is still available irrespective of the location of a fire.
Approved Document B (ADB)[24]	Guidance issued by Government in support of the fire safety aspects of the building regulations.
As low as reasonably practicable	Is a concept where risks should continue to be reduced until you reach a point where the cost and effort to reduce the risk further would be grossly disproportionate to the benefit achieved.
Automatic fire detection system	A means of automatically detecting the products of a fire and sending a signal to a fire warning system. See 'Fire warning'.
Basement	A storey with a floor which at some point is more than 1,200mm below the highest level of ground adjacent to the outside walls, unless, and for escape purposes only, such area has adequate, independent and separate means of escape.
Child	Anyone who is not over compulsory school age, i.e. before or just after their 16th birthday.
Class 0, 1 or 3 surface spread of flame	Classes of surface spread of flame for materials needed to line the walls and ceilings of escape routes. See Appendix B for further information.
Combustible material	A substance that can be burned.
Compartment wall and/or floor	A fire-resisting wall or floor that separates one fire compartment from another.

Term	Definition
Competent person	A person with enough training and experience or knowledge and other qualities to enable them properly to assist in undertaking the preventive and protective measures.
Dangerous substance	1. A substance which because of its physico-chemical or chemical properties and the way it is used or is present at the workplace creates a risk.
	2. A substance subject to the Dangerous Substances and Explosive Atmosphere Regulations 2002 (DSEAR).
Dead end	Area from which escape is possible in one direction only.
Direct distance	The shortest distance from any point within the floor area to the nearest storey exit, or fire-resisting route, ignoring walls, partitions and fixings.
Domestic premises	Premises occupied as a private dwelling, excluding those areas used in common by the occupants of more than one such dwelling.
Emergency escape lighting	Lighting provided to illuminate escape routes that will function if the normal lighting fails.
Enforcing authority	The fire and rescue authority or any other authority specified in Article 25 of the Regulatory Reform (Fire Safety) Order 2005[1].
Escape route	Route forming that part of the means of escape from any point in the premises to a final exit.
Evacuation lift	A lift that may be used for the evacuation of people with disabilities, or others, in a fire.
External escape stair	Stair providing an escape route, external to the building.
Fail-safe	Locking an output device with the application of power and having the device unlock when the power is removed. Also known as fail unlock, reverse action or power locked.
False alarm	A fire signal, usually from a fire warning system, resulting from a cause other than fire.
Final exit	An exit from a building where people can continue to disperse in safety and where they are no longer at danger from fire and/or smoke.
Fire compartment	A building, or part of a building, constructed to prevent the spread of fire to or from another part of the same building or an adjoining building.
Fire door	A door or shutter, together with its frame and furniture, provided for the passage of people, air or goods which, when closed is intended to restrict the passage of fire and/or smoke to a predictable level of performance.

Term	Definition
Firefighting lift	A lift, designed to have additional protection, with controls that enable it to be used under the direct control of the fire and rescue service when fighting a fire.
Firefighting shaft	A fire-resisting enclosure containing a firefighting stair, fire mains, firefighting lobbies and if provided, a firefighting lift.
Firefighting stairway	See firefighting shaft.
Fire resistance	The ability of a component or construction of a building to satisfy, for a stated period of time, some or all of the appropriate criteria of relevant standards. (Generally described as 30 minutes fire-resisting or 60 minutes fire-resisting.) See BS EN 1363-1,[46] BS 476-7[33] and associated standards for further information.
Fire safety manager	A nominated person with responsibility for carrying out day-to-day management of fire safety. (This may or may not be the same as the 'responsible person'.)
Fire safety strategy	A number of planned and co-ordinated arrangements designed to reduce the risk of fire and to ensure the safety of people if there is a fire.
Fire stopping	A seal provided to close an imperfection of fit or design tolerance between elements or components, to restrict the passage of fire and smoke.
Fire-warning system	A means of alerting people to the existence of a fire. (See automatic fire detection system.)
Flammable material	Easily ignited and capable of burning rapidly.
Highly flammable	Generally liquids with a flashpoint of below 21°C. (The Chemicals Hazard Information and Packaging for Supply Regulations 2002[47] (CHIP) give more detailed guidance.)
Hazardous substance	1. See Dangerous substance. 2. A substance subject to the Control of Substances Hazardous to Health Regulations 2002 (COSHH).
Inner room	A room from which escape is possible only by passing through another room (the access room).
Licensed premises	Any premises that require a licence under any statute to undertake trade or conduct business activities.
Material change	An alteration to the premises, process or service which significantly affects the level of risk to people from fire in those premises.

Term	Definition
Means of escape	Route(s) provided to ensure safe egress from the premises or other locations to a place of total safety.
Phased evacuation	A system of evacuation in which different parts of the premises are evacuated in a controlled sequence of phases, those parts of the premises expected to be at greatest risk being evacuated first.
Place of reasonable safety	A place within a building or structure where, for a limited period of time, people will have some protection from the effects of fire and smoke. This place, usually a corridor or stairway, will normally have a minimum of 30 minutes fire resistance and allow people to continue their escape to a place of total safety.
Place of total safety	A place, away from the premises, in which people are at no immediate danger from the effects of a fire.
Premises	Any place, such as a building and the immediate land bounded by any enclosure of it, any tent, moveable or temporary structure or any installation or workplace.
Protected lobby	A fire-resisting enclosure providing access to an escape stairway via two sets of fire doors and into which no room opens other than toilets and lifts.
Protected stairway	A stairway which is adequately protected from the rest of the building by fire-resisting construction.
Protected route	An escape route which is adequately protected from the rest of the building by a fire-resisting construction.
Refuge	A place of reasonable safety in which a disabled person and others who may need assistance may rest or wait for assistance before reaching a place of total safety. It should lead directly to a fire-resisting escape route.
Responsible person	The person ultimately responsible for fire safety as defined in the Regulatory Reform (Fire Safety) Order 2005.[1]
Relevant persons	Any person lawfully on the premises and any person in the immediate vicinity, but does not include firefighters carrying out firefighting duties.
Self-closing device	A device that is capable of closing the door from any angle and against any latch fitted to the door.
Significant finding	A feature of the premises, from which the fire hazards and persons at risk are identified. The actions you have taken or will take to remove or reduce the chance of a fire occuring or the spread of fire and smoke. The actions people need to take in case of fire. The necessary information, instruction and training needed and how it will be given

Term	Definition
Smoke alarm	Device containing within one housing all the components, except possibly the energy source, for detecting smoke and giving an audible alarm.
Staged fire alarms	A fire warning which can be given in two or more stages for different purposes within a given area (i.e. notifying staff, stand by to evacuate, full evacuation).
Storey exit	A final exit or a doorway giving direct access into a protected stairway, firefighting lobby, or external escape route.
Travel distance	The actual distance to be travelled by a person from any point within the floor area to the nearest storey exit or final exit, having regard to the layout of walls, partitions and fixings.
Vision panel	A transparent panel in a wall or door of an inner room enabling the occupant to become aware of a fire in the access area during the early stages.
Way guidance	Low mounted luminous tracks positioned on escape routes in combination with exit indicators, exit marking and intermediate direction indicators along the route, provided for use when the supply to the normal lighting fails, which do not rely on an electrical supply for their luminous output.
Where necessary	The Order requires that fire precautions (such as firefighting equipment, fire detection and warning, and emergency routes and exits) should be provided (and maintained) 'where necessary'. What this means is that the fire precautions you must provide (and maintain) are those which are needed to reasonably protect relevant persons from risks to them in case of fire. This will be determined by the findings of your risk assessment including the preventative measures you have or will have taken. In practice, it is very unlikely, that a properly conducted fire risk assessment, which takes into account all the matters relevant for the safety of persons in case of fire, will conclude that no fire precautions (including maintenance) are necessary.
Young person	(a) A person aged 16 years, from the date on which he attains that age until and including the 31st August which next follows that date. (b) A person aged 16 years and over who is undertaking a course of full-time education at a school or college which is not advanced education. (c) A person aged 16 years and over who is undertaking approved training that is not provided through a contract of employment. For the purposes of paragraphs (b) and (c) the person: (a) shall have commenced the course of full-time education or approved training before attaining the age of 19 years; and (b) shall not have attained the age of 20 years.

References

The following documents are referenced in this guide. Where dated only this version applies. Where updated, the latest version of the document applies.

1 Regulatory Reform (Fire Safety) Order 2005, SI 2005/1541. The Stationery Office, 2005. ISBN 0 11 072945 5.

2 Fire Precautions Act 1971 (c 40). The Stationery Office, 1971. ISBN 0 10 544071 X.

3 Fire Precautions (Workplace) Regulations 1997, SI 1997/1840. The Stationery Office, 1997. ISBN 0 11 064738 6.

4 Fire Precautions (Workplace) (Amendment) Regulations 1999, SI 1999/1877. The Stationery Office, 1999. ISBN 0 11 082882 8.

5 Health and Safety (Safety Signs and Signals) Regulations 1996, SI 1996/341. The Stationery Office, 1996. ISBN 0 11 054093 X.

6 *Safety signs and signals. The Health and Safety (Safety Signs and Signals) Regulations 1996. Guidance on regulations,* L64. HSE Books, 1996. ISBN 0 7176 0870 0.

7 Dangerous Substances and Explosive Atmospheres Regulations 2002, SI 2002/2776. The Stationery Office, 2002. ISBN 0 11 042957 5.

8 *Dangerous substances and explosive atmospheres. Dangerous Substances and Explosive Atmospheres Regulations 2002. Approved code of practice and guidance,* L138. HSE Books, 2003. ISBN 0 7176 2203 7.

9 *Storage of full and empty LPG cylinders and cartridges. Code of Practice 7.* LP Gas Association 2000. Available from LP Gas Association, Pavilion 16, Headlands Business Park, Salisbury Road, Ringwood, Hampshire BH24 3PB.

10 *Maintaining portable electrical equipment in offices and other low-risk environments,* INDG236. HSE Books, 1996. (ISBN 0 7176 1272 4 single copy free or priced packs of 10.)

11 Construction (Health, Safety and Welfare) Regulations 1996, SI 1996/1592. The Stationery Office, 1996. ISBN 0 11 035904 6.

12 *A guide to the Construction (Health, Safety and Welfare) Regulations 1996,* INDG220. HSE Books, 1996. (ISBN 0 7176 1161 2 single copy free or priced packs of 10.)

Health and safety in construction, HSG150 (second edition). HSE Books, 2001. ISBN 0 7176 2106 5.

13 Disability Discrimination Act 1995 (c 50). The Stationery Office, 1995. ISBN 0 10 545095 2.

14 BS 8300: *The design of buildings and their approaches to meet the needs of disabled people. Code of practice.* British Standards Institution. ISBN 0 580 38438 1.

15 ODPM/CACFOA/BFPSA guidance on reducing false alarms.

16 BS 5839-1: *Fire detection and alarm systems for buildings. Code of practice for system design, installation, commissioning and maintenance,* British Standards Institution. ISBN 0 580 40376 9.

17 Manual Handling Operations Regulations 1992, SI 1992/2793. The Stationery Office, 1992. ISBN 0 11 025920 3.

18 BS 5306-8: *Fire extinguishing installations and equipment on premises. Selection and installation of portable fire extinguishers. Code of practice.* British Standards Institution. ISBN 0 580 33203 9.

19 BS 5306-3: *Fire extinguishing installations and equipment on premises. Code of practice for the inspection and maintenance of portable fire extinguishers.* British Standards Institution. ISBN 0 580 42865 6.

20 BS 7863: *Recommendations for colour coding to indicate the extinguishing media contained in portable fire extinguishers.* British Standards Institution. ISBN 0 580 25845 9.

21 BS EN 671-3: *Fixed firefighting systems. Hose systems. Maintenance of hose reels with semi-rigid hose and hose systems with lay-flat hose*. British Standards Institution. ISBN 0 580 34112 7.

22 BS EN 12845: *Fixed firefighting systems. Automatic sprinkler systems. Design, installation and maintenance*. British Standards Institution. ISBN 0 580 44770 7.

23 Workplace (Health, Safety and Welfare) Regulations 1992, SI 1992/3004. The Stationery Office, 1992. ISBN 0 11 025804 5.

24 *The Building Regulations 2000: Approved Document B fire safety*. The Stationery Office. ISBN 0 11 753911 2.

25 Local Government (Miscellaneous Provisions) Act 1982 (c 30). The Stationery Office, 1982. ISBN 0 10 543082 X.

26 BS 5395-2: *Stairs, ladders and walkways. Code of practice for the design of industrial type stairs, permanent ladders and walkways*. British Standards Institution. ISBN 0 580 14706 1.

27 BS 5266-8: *Emergency lighting. Code of practice for emergency escape lighting systems*. British Standards Institution.

28 BS 5266-1: *Emergency lighting. Code of practice for the emergency lighting of premises*. British Standards Institution.

29 BS EN 1838: *Lighting applications. Emergency lighting*. British Standards Institution. ISBN 0 580 32992 5.

30 BS 5499-4: *Safety signs, including fire safety signs. Code of practice for escape route signing*. British Standards Institution.

31 BS 7974: *Application of fire safety engineering principles to the design of buildings. Code of practice*. British Standards Institution. ISBN 0 580 38447 0.

32 *Successful health and safety management,* HSG65 (second edition). HSE Books, 1997. ISBN 0 7176 1276 7.

33 BS 476-7: *Fire tests on building materials and structures. Method of test to determine the classification of the surface spread of flame of products*. British Standards Institution.

34 BS EN 13501-1: *Fire classification of construction products and building elements. Classification using test data from reaction to fire tests*. British Standards Institution.

35 *Guidelines for the construction of fire resisting structural elements,* BR 128. Building Research Establishment, 1988.

36 BS EN 1634-1: *Fire resistance tests for door and shutter assemblies. Fire doors and shutters*. British Standards Institution. ISBN 0 580 32429 X.

37 BS 476-22: *Fire tests on building materials and structures. Methods for determination of the fire resistance of non-loadbearing elements of construction*. British Standards Institution. ISBN 0 580 15872 1.

38 BS EN 1935: *Building hardware. Single-axis hinges. Requirements and test methods*. British Standards Institution. ISBN 0 580 39272 4.

39 BS EN 1154: *Building hardware. Controlled door closing devices. Requirements and test methods*. British Standards Institution. ISBN 0 580 27476 4.

40 BS EN 1634-2: *Fire resistance tests for door and shutter assemblies. Part 2. Fire door hardware. Building hardware for fire resisting doorsets and openable windows*. British Standards Institution.

41 BS 5839-3: *Fire detection and alarm systems for buildings. Specification for automatic release mechanisms for certain fire protection equipment*. British Standards Institution. ISBN 0 580 15787 3.

42 BS 8214: *Code of practice for fire door assemblies with non-metallic leaves*. British Standards Institution. ISBN 0 580 18871 6.

43 BS EN 1125: *Building hardware. Panic exit devices operated by a horizontal bar. Requirements and test methods*. British Standards Institution.

44 BS EN 179: *Building hardware. Emergency exit devices operated by a lever handle or push pad. Requirements and test methods*. British Standards Institution. ISBN 0 580 28863 3.

45 BS 8220: *Guide for security of buildings against crime.* British Standards Institution. ISBN 0 580 23692 7.

46 BS EN 1363-1: *Fire resistance tests. General requirements.* British Standards Institution.

47 Chemicals (Hazard Information and Packaging for Supply) Regulations 2002, SI 2002/1689. The Stationery Office, 2002. ISBN 0 11 042419 0.

48 *Guide to the safe warehousing of aerosols.* British Aerosol Manufacturers Association.

49 *The storage of flammable liquids in containers,* HSG51. Health and Safety Executive.

50 *Manufacture and storage of explosives. Manufacture and Storage of Explosives Regulations 2005. Approved code of practice and guidance,* L139. HSE Books, 2005. ISBN 0 7176 2816 7.

51 *Storing and selling fireworks safely,* INDG407. Health and Safety Executive.

52 BS 5588-12: *Fire Precautions in the design, construction and use of buildings – Part 12: Managing fire safety.* British Standards Institution.

53 The Electricity at Work Regulations 1989, SI 1989/635.

54 The Electrical Equipment (Safety) Regulations 1994, SI 1994/3260.

55 The Construction (Design and Management) Regulations 1994 (CONDAM/CDM Regs). HMSO, 1994.

56 *Construction Information Sheet No. 51: Construction fire safety.* Health and Safety Executive.

57 *Fire safety in construction work, HSG 168.* Health and Safety Executive. ISBN 0 7176 1332 1.

58 *Fire prevention on construction sites: The joint code of practice on the protection from fire of construction sites and buildings undergoing renovation* (fifth edition). Fire Protection Association and Construction Federation, 2000. ISBN 0 90216739 1.

59 *Design, construction, specification and fire management of insulated envelopes for temperature controlled environments.* International Association of Cold Storage Contractors.

60 BS 5306-2: *Fire extinguishing installations and equipment on premises. Specification for sprinkler systems.* British Standards Institution.

61 BS 5588-5: *Fire precautions in the design, construction and use of buildings. Access and facilities for firefighting.* British Standards Institution.

62 *The Building Regulations 1991: Approved Document M access to and use of buildings* (2004 edition). The Stationery Office.

63 BS 5588-8: *Fire precautions in the design, construction and use of buildings. Code of practice for means of escape for disabled people.* British Standards Institution.

64 *CIBSE Guide Volume E: Fire engineering.* Chartered Institution of Building Services Engineers, 1997.

65 *Design methodologies for smoke and heat exhaust ventilation,* Report 368, BRE, 1999.

66 *A Guide to best practice in the specification and use of rire-resistant glazed systems.* Glass and Glazing Federation, 2005.

67 *Increasing the fire resistance of existing timber doors,* Information Paper 8/82. BRE.

68 *Fire resisting doorsets by upgrading,* Wood Information Sheet 1-32. Timber Research and Development Association.

69 BS 4787-1: *Internal and external wood doorsets, door leaves and frames. Specification for dimensional requirements.* British Standards Institution.

70 BS EN 1155: *Building hardware. Electrically powered hold-open devices for swing doors. Requirements and test methods.* British Standards Institution.

71 BS EN 1158: *Building hardware. Door coordinator devices. Requirements and test methods.* British Standards Institution.

72 *Hardware for timber fire and escape doors.* BHIF, 2000.

73 BS 5266-6: *Emergency lighting. Code of practice for non-electrical low mounted way guidance systems for emergency use. Photoluminescent systems.* British Standards Institution.

74 BS 5499-5: *Graphical symbols and signs. Safety signs, including fire safety signs. Signs with specific safety meanings.* British Standards Institution.

75 'Sprinklers: High piled and rack storage'. *Fire Surveyor, Vol. 9, No. 1, 21–26, February 1990.*

76 *Safe handling of combustible dusts: Precautions against explosions,* HSG103 (second edition). HSE Books, 2003. ISBN 0 7176 2726 8.

77 BS EN 12209: Building hardware. Locks and latches. *Mechanically operated locks, latches and locking plate.* British Standards Institution.

78 Licensing Act 2003. The Stationery Office. ISBN 0 10 541703 3.

79 BS EN 3-7: *Portable fire extinguishers. Characteristics, performance requirement and test methods.* British Standards Institution.

80 BS EN 1906: *Building hardware. Lever handles and knob furniture. Requirements and test methods.* British Standards Institution.

Further reading

The latest versions of all documents listed in this section should be used, including any amendments.

Any views expressed in these documents are not necessarily those of the DCLG.

BS 4422	Fire. Vocabulary. British Standards Institution.
BS PD 6512-3	Use of elements of structural fire protection with particular reference to the recommendations given in BS 5588 *Fire precautions in the design and construction of buildings. Guide to the fire performance of glass.* British Standards Institution.
BS EN 81	Safety rules for the construction and installation of lifts. British Standards Institution.
BS EN 81-70	Safety rules for the construction and installation of lifts. Particular applications for passenger and goods passenger lifts. Accessibility to lifts for persons including persons with disability. British Standards Institution.
BS 5041-1	Fire hydrant systems equipment. Specification for landing valves for wet risers. British Standards Institution.
BS 5041-2	Fire hydrant systems equipment. Specification for landing valves for dry risers. British Standards Institution.
BS 5041-3	Fire hydrant systems equipment. Specification for inlet breechings for dry riser inlets. British Standards Institution.
BS 5041-4	Fire hydrant systems equipment. Specification for boxes for landing valves for dry risers. British Standards Institution.
BS 5041-5	Fire hydrant systems equipment. Specification for boxes for foam inlets and dry riser inlets. British Standards Institution.
BS 9990	Code of practice for non-automatic firefighting systems in buildings. British Standards Institution.
BS 7944	Type 1 heavy duty fire blankets and type 2 heavy duty heat protective blankets. British Standards Institution.
BS EN 1869	Fire blankets. British Standards Institution.
BS ISO 14520-1	Gaseous fire-extinguishing systems. Physical properties and system design. General requirements. British Standards Institution.
BS 5266-2	Emergency lighting. Code of practice for electrical low mounted way guidance systems for emergency use. British Standards Institution.
BS EN 60598-1	Luminaires. General requirements and tests. British Standards Institution.

BS 5499-1	Graphical symbols and signs. Safety signs, including fire safety signs. Specification for geometric shapes, colours and layout. British Standards Institution.
BS EN 1634-3	Fire resistance tests for door and shutter assemblies. Smoke control doors and shutters. British Standards Institution.
Draft BS EN 14637	Building hardware. Electrically controlled hold-open systems for fire/smoke door assemblies. Requirements, test methods, application and maintenance. (Consultation document.) British Standards Institution.
BS EN 45020	Standardisation and related activities. General vocabulary. British Standards Institution.
ISO 13784-2	Reaction to fire tests for sandwich panel building systems. Part 2: test method for large rooms. British Standards Institution.
BS 5268-4.2	Structural use of timber. Fire resistance of timber structures. Recommendations for calculating fire resistance of timber stud walls and joisted floor constructions. British Standards Institution.
BS 5588-11	Fire precautions in design, construction and use of buildings. Part 11: Code of practice for shops, offices, industrial, storage and other similar buildings. British Standards Institution.

Design principles of fire safety. The Stationery Office, 1996. ISBN 0 11 753045 X.

Chemicals (Hazard Information and Packaging for Supply) Regulations 2002, SI 2002/1689. The Stationery Office, 2002. ISBN 0 11 042419 0. Supporting guides: *The idiot's guide to CHIP 3: Chemicals (Hazard Information and Packaging for Supply) Regulations 2002,* INDG350. HSE Books, 2002. (ISBN 0 7176 2333 5 single copy free or priced packs of 5); *CHIP for everyone,* HSG228. HSE Books, 2002. ISBN 0 7176 2370 X.

Guidance on the acceptance of electronic locks to doors required for means of escape. The Chief and Assistant Chief Fire Officers' Association.

Ensuring best practice for passive fire protection in buildings. Building Research Establishment, 2003. ISBN 1 870409 19 1

Smoke shafts protecting fire shafts: their performance and design, BRE Project Report 79204. Building Research Establishment, 2002.

Fire safety of PTFE-based material used in building, BRE Report 274. Building Research Establishment, 1994. ISBN 0 851256 53 8.

Fires and human behaviour. David Fulton Publishers, 2000. ISBN 1 85346 105 9

Management of health and safety at work. Management of Health and Safety at Work Regulations 1999. Approved code of practice and guidance, L21 (second edition). HSE Books, 2000. ISBN 0 7176 2488 9.

Fire safety in construction work, HSG168. HSE Books, 1997. ISBN 0 7176 1332 1.

LPC rules for automatic sprinkler installations. The Fire Protection Association, 2003.

Index

Page numbers in italics refer to information in Figures or Tables.